# Corrosion Protection for the Oil and Gas Industry

## Pipelines, Subsea Equipment, and Structures

# Corrosion Protection for the Oil and Gas Industry

## Pipelines, Subsea Equipment, and Structures

Authored by
Mavis Sika Okyere

CRC Press
Taylor & Francis Group
Boca Raton London New York

CRC Press is an imprint of the
Taylor & Francis Group, an **informa** business

CRC Press
Taylor & Francis Group
6000 Broken Sound Parkway NW, Suite 300
Boca Raton, FL 33487-2742

First issued in paperback 2020

ISBN 13: 978-0-367-65653-9 (pbk)
ISBN 13: 978-0-367-17280-0 (hbk)

---

**Library of Congress Cataloging-in-Publication Data**

---

Names: Okyere, Mavis Sika, author.
Title: Corrosion protection for the oil and gas industry : pipelines, subsea equipment, and structures / Mavis Sika Okyere.
Description: Boca Raton : Taylor & Francis, a CRC title, part of the Taylor & Francis imprint, a member of the Taylor & Francis Group, the academic division of T&F Informa, plc, 2018. | Includes bibliographical references and index.
Identifiers: LCCN 2018048427| ISBN 9780367172800 (hardback : acid-free paper) | ISBN 9780429056451 (ebook)
Subjects: LCSH: Offshore structures--Corrosion. | Offshore structures--Protection. | Offshore oil well drilling--Equipment and supplies--Protection. | Corrosion and anti-corrosives.
Classification: LCC TC1670 .O48 2018 | DDC 622/.33819--dc23
LC record available at https://lccn.loc.gov/2018048427

---

**Visit the Taylor & Francis Web site at**
**http://www.taylorandfrancis.com**

**and the CRC Press Web site at**
**http://www.crcpress.com**

# Contents

# Preface

On average, oil and gas companies, and water/wastewater industries, use six percent of their annual income to combat corrosion.

This book has been prepared to satisfy the needs of students, practicing engineers, and scientists for an introduction to corrosion protection for the oil and gas industry and to the task of overcoming corrosion issues. I have made the subject very concise, and the book is easily understood. It is concerned primarily with the external and internal corrosion protection of onshore pipelines and subsea pipelines, but reference is also made to the protection of other subsea equipment, subsea structures, risers, and shore approaches. The corrosion protection of pipelines (either onshore or subsea), which operate at elevated temperatures and therefore require the application of an anticorrosion and/or thermal insulation coating capable of withstanding such corrosion is also covered.

Two case studies, design examples, and my experience as a pipeline integrity engineer are featured in this book. Readers should gain a high-quality and in-depth understanding of the corrosion protection methods available and be able to apply them to solve corrosion engineering problems.

Education and training in corrosion protection methods was the main concern and motivation behind the production of this book. I hope that it will be of substantial benefit to nations, to various industries, and to individuals all over the world who will take advantage of the opportunity afforded them by this educational effort. I would like to express my gratitude to all the reviewers who have helped to improve the book.

**Mavis Sika Okyere**
*Ghana National Gas Company Ltd.*

# Acknowledgments

I am grateful to God for the good health and well-being that were needed to write this book.

The completion of this book could not have been possible without the participation and assistance of those who offered comments and assisted in editing and proofreading. So many other people helped as well, and their advice and contributions are appreciated and gratefully acknowledged.

I am grateful to my parents and siblings for their immense love and care. They have always encouraged me to explore my potential and pursue my dreams.

To my children and my husband, all of whom in one way or another shared their support, either morally, financially, or physically—thank you. May God bless all of you!

# Author

**Mavis Sika Okyere** is a pipeline integrity engineer at Ghana National Gas Company. She is an expert in risk-based assessment, pipeline integrity, corrosion control, and cathodic protection design. She has experience with subsea structural engineering, piping, and pipeline engineering principles as applied to both onshore and offshore conditions.

Mavis holds an MSc. in gas engineering and management from the University of Salford, United Kingdom, and a BSc. in civil engineering from Kwame Nkrumah University of Science and Technology, Ghana.

She has over four years of experience in pipeline integrity engineering. In her current role at Ghana National Gas Company, she is involved in coating, cathodic protection, key performance indicators, leak detection, atmospheric corrosion survey, corrosion monitoring, and pipe-to-soil potential measurement.

In addition, her industry experience includes teaching oil and gas management courses at Bluecrest College, serving as an engineer with LUDA Development Ltd, INTECSEA/Worleyparsons Atlantic Ltd, Technip, Ussuya Ghana Ltd, and the Ghana Highway Authority. She has published in several books and in various journals and is a member of many national and international bodies such as NACE International.

# 1 Introduction

A corrosion protection method is a technique used to minimize corrosion, such as the application of an anticorrosion coating, cathodic protection, or other methods that make metal resistant to corrosion. Corrosion protection methods are also known as corrosion control.

Corrosion causes hundreds of billions of dollars in losses each year. Corrosion is a very common yet serious problem across different industries and environments. Corrosion can be increased by the presence of high temperatures, high relative humidity, salt air, harsh chemicals, and even by mold, moss, and dirt deposits. A 2016 study by the National Association of Corrosion Engineers (NACE) pegged the global cost of corrosion at $2.5 trillion. Corrosion is not simply a sustainment concern; it needs to be addressed from program/system/equipment inception to disposal. Corrosion prevention and control at the early stages in a pipeline system development can result in a reduction of total ownership cost.

Five primary techniques used for limiting corrosion rates to practical levels are:

- Material selection
- Coatings
- Cathodic protection
- Chemical injection
- Proper anticorrosion design

The content of this book is developed from broad and in-depth experience regarding the protection of pipeline systems/equipment from corrosion and its effects. The book is arranged in the following sections:

- A summary of the main causes of corrosion and the requirements for protection of materials against corrosion.
- A summary of the selection of corrosion-resistant materials.
- A summary of coating materials commonly used for corrosion protection and the limitations to their use, their application, and repair.
- Guidelines for the design of cathodic protection systems and reviews of cathodic protection methods, materials, installation, and monitoring.
- Galvanic zinc application.
- Chemical corrosion inhibitor.
- Case studies and design example.
- A reference section detailing source documents and support documents which may be used in conjunction with this book.

It should be noted that guidance in this book is applicable to most locations around the world. However, there may be local constraints on corrosion coating selection, for example, which are not immediately evident if the design is performed from a distance.

This book is concerned with the protection of steels against corrosion in an aqueous environment, either immersed in seawater or buried. For most oil and gas applications, the steels are of the carbon or low alloy types, although in some cases, it is necessary to protect stainless steels.

# 2 Corrosion

Corrosion is the deterioration of a material (usually metal) due to its interaction with the environment. Corrosion is a natural phenomenon, which should not surprise one, but rather should be expected to occur. Metals are high-energy materials which exist because heat energy was added to natural iron ores during the smelting process. Nature, by environmental contact, constantly attacks these high-energy materials and breaks them down to the natural elements from which they were derived.

The effect of corrosion on oil and gas pipelines can be catastrophic. The destruction of the metal eventually leads to leakages, which have the potential to cause massive disasters such as fires and explosions. High safety concerns and strict standards applicable to the oil and gas industry mean that constant monitoring is needed to identify the presence and extent of corrosion.

## 2.1 BASICS OF AQUEOUS METALLIC CORROSION

A corroding system is driven by two spontaneous reactions that take place at the interface between the metal and an aqueous environment. The two simultaneous reactions are the oxidation (anodic) reaction and a reduction (or cathodic reaction). The first reaction is when the chemical species from the aqueous environment remove electrons from the metal; the other is a reaction in which metal surface atoms participate to replenish the electron deficiency.

Example of anodic (oxidation, electron donating) reaction:

$$Fe \rightarrow Fe^{2+} + 2e^{-} \left(metal\,dissolution\right)$$

Examples of cathodic (reduction, electron accepting) reactions:

$$O_2 + 2H_2O + 4e^{-} \rightarrow 4OH^{-} \left(oxygen\,reduction\right)$$

$$2H^{+} + 2e^{-} \rightarrow H_2 \left(hydrogen\,evolution\right)$$

An important effect is the electron exchange between the two reactions that constitutes an electronic current at the metal surface. This, in turn, imposes an electric potential on the metal surface of such a value that the supply and demand for electrons in the two coupled reactions are balanced.

The potential imposed on the metal is of much greater significance than simply to balance the complementary reactions which produce it. This is because it is one of the principal factors determining what the reactions will be.

Generally, at the anodic locations, there is corrosion damage (e.g., metal loss), while at the cathodic location, no corrosion damage occurs. The location of anodes and cathodes tends to move randomly over the surface of the metal for alloys that

3

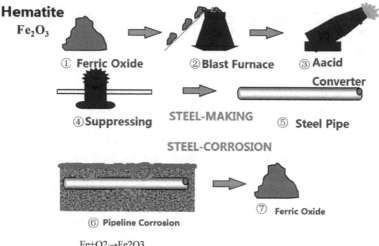

**FIGURE 2.1**    Basics of corrosion. (Source: Tong, Shan. 2015. *Cathodic protection*. Training document, Ghana: Sinopec.)

are subject to general corrosion. However, the location of an anode tends to become strongly localized for corrosion-resistant alloys, which are covered by a passive oxide film. This gives rise to localized corrosion damage such as pitting corrosion, stress corrosion cracking, and crevice corrosion (Figures 2.1–2.2).

## 2.2   FORMS OF CORROSION

Corrosion is generally divided into general corrosion and localized corrosion. Localized corrosion is classified as follows:

1. Microbiological corrosion
2. Galvanic corrosion
3. Crevice corrosion
4. Pitting corrosion
5. Galvanic corrosion
6. Erosion corrosion
7. Stress corrosion cracking (SCC)
8. Fatigue corrosion

This classification of corrosion is based on visual characteristics of morphology of attack as well as the type of environment to which the surface is exposed.

Microbiological corrosion, also called bacterial corrosion, bio-corrosion, microbiologically influenced corrosion, or microbially induced corrosion (MIC), is corrosion caused or promoted by microorganisms. It can apply to both metals and nonmetallic materials.

Crevice corrosion is a localized attack on a metal adjacent to the crevice between two joining surfaces (two metals or metal-nonmetal crevices). The corrosion is generally confined to one localized area of one metal (Natarajan, 2012). This type of corrosion can be initiated by concentration gradients (due to ions or oxygen). Accumulation of chlorides inside the crevice will aggravate damage (see Figure 2.2). Various factors influence crevice corrosion, such as:

- Materials: alloy composition, metallographic structure
- Environmental conditions such as pH, oxygen concentration, halide concentrations, temperature
- Geometrical features of crevices, surface roughness
- Metal-to-metal or metal-to-nonmetal type

Filiform corrosion is a special type of crevice corrosion.

Pitting corrosion is a localized phenomenon confined to smaller areas. Formation of micro-pits can be very damaging. Pitting factor (ratio of deepest pit to average penetration) can be used to evaluate the severity of pitting corrosion which is usually observed in passive metals and alloys. Concentration cells involving oxygen gradients or ion gradients can initiate pitting through generation of anodic and cathodic areas. Chloride ions are damaging to the passive films and can make pit formation autocatalytic. Pitting tendency can be predicted through measurement of pitting potentials. Similarly, critical pitting temperature is also a useful parameter.

Galvanic corrosion often referred to as dissimilar metal corrosion occurs in galvanic couples where the active one corrodes. EMF series (thermodynamic) and galvanic series (kinetic) could be used for prediction of this type of corrosion. Galvanic corrosion can occur in multiphase alloys (Natarajan, 2012) (Figure 2.3).

Erosion corrosion is the deterioration of metals and alloys due to relative movement between surfaces and corrosive fluids. Depending on the rate of this movement,

**FIGURE 2.2** Crevice corrosion. (Source: https://www.corrosionpedia.com/whats-the-inside-scoop-on-crevice-corrosion/2/6588.)

**FIGURE 2.3** Galvanic corrosion. (Source: https://www.corrosionpedia.com/21-types-of-pipe-corrosion-failure/2/1484.)

abrasion takes place. This type of corrosion is characterized by grooves and surface patterns having directionality. Typical examples are:

- Stainless alloy pump impeller
- Condenser tube walls

All equipment types exposed to moving fluids are prone to erosion corrosion. Many failures can be attributed to impingement (impingement attack). Erosion corrosion due to high-velocity impingement occurs in steam condenser tubes, slide valves in petroleum refinery at high temperature, inlet pipes, cyclones, and steam turbine blades. Cavitation damage can be classified as a special form of erosion corrosion.

Stress corrosion cracking (SCC) refers to failure under the simultaneous presence of a corrosive medium and tensile stress. Two classic examples of SCC are caustic embrittlement of steels occurring in riveted boilers of steam-driven locomotives and season cracking of brasses observed in brass cartridge cases due to ammonia in the environment. Stress cracking of different alloys does occur depending on the type of corrosive environment. Stainless steels crack in a chloride atmosphere. Major variables influencing SCC include solution composition, metal/alloy composition and structure, stress, and temperature (Natarajan, 2012).

The causes of Stress Corrosion Cracking (SCC) are:

1. A Compelling Cracking Environment: The four factors controlling the formation of the potent environment for the initiation of SCC are the type and condition of the coating, soil, temperature, and cathodic current levels.
   - Pipeline Coating Types: SCC normally starts on the pipeline surface at areas where coating disbondment or coating damage occurs. The resistance of a coating to disbonding is a primary performance property that affects all forms of external pipeline corrosion.
   - Soil: The soil type, including drainage, carbon dioxide ($CO_2$), temperature, moisture content of the soil, and electrical conductivity, forms a conducive environment for the formation of stress corrosion cracks.

- Cathodic Protection: Cathodic protection (CP) current forms a carbonate/bicarbonate environment at the pipeline surface, where high pH SCC occurs. For near-neutral pH, SCC CP is absent.
- Temperature: Temperature has a significant effect on the occurrence of high pH SCC, while it has no effect on near-neutral pH SCC.
2. Pipe material characteristics: A number of pipe characteristics and qualities, such as the pipe manufacturing process, type of steel, grade of steel, cleanliness of the steel (presence or absence of impurities or inclusions), steel composition, plastic deformation characteristics of the steel (cyclic-softening characteristics), steel temperature and pipe surface condition, are necessary conditions for the formation of stress corrosion cracks (SCC).
3. When the tensile stress is greater than the threshold stress: If the tensile stress is higher than the threshold stress, this can lead to SCC, especially when there is some dynamic or cyclic component to the stress.

## 2.3  POLARIZATION AND CORROSION RATES

### 2.3.1  CONCEPT OF POLARIZATION

Polarization refers to electrode potential variation caused by current flowing through the electrode, including anodic polarization and cathode polarization. No matter anodic polarization or cathode polarization can reduce potential difference between two electrodes of corrosion cell, reducing corrosion current and preventing the corrosion process.

Essentially, polarization reflects suppression of electrode process. Polarization degree reflects resistance degree, while over-potential of electrode reaction reflects impetus of this electrode reaction.

#### 2.3.1.1  Causes of Cathodic Polarization

Cathodic process is a reduction process in which the substance that can absorb electrons in the solution obtains electrons. Causes of its potential becoming negative and polarized are not unitary.

1. Electrochemical polarization: Usually, the speed of the reduction reaction combining oxidant and electrons at the cathode is slower than the speed of electrons flowing from anode, resulting in electron accumulation and potential becoming negative, which is called electrochemical polarization.
2. Concentration polarization: Concentration polarization happens when hydrogen ions generated at the anode are attracted to the cathode and the hydroxide ions generated at the cathode are attracted to the anode.

#### 2.3.1.2  Polarization Diagram

The polarization diagram is used to analyze the main controlling factor of the corrosion process, illustrating the physical meaning of the corrosion potential of different corrosion systems. If the corrosion potential is close to the balance potential of the local anode reaction, the corrosion process is determined by cathodic polarization. Meanwhile, self-corrosion current is determined by cathode polarization, as shown in Figure 2.4, and the corrosion current is decided by anode polarization, as shown in Figure 2.5.

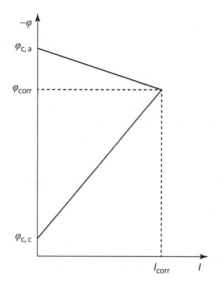

FIGURE 2.4   Cathodic polarization control. (Source: Tong, Shan. 2015. *Cathodic protection*. Training document, Ghana: Sinopec.)

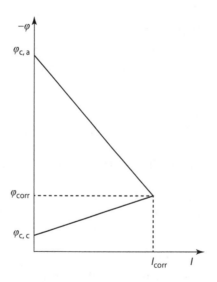

FIGURE 2.5   Anode polarization control. (Source: Tong, Shan. 2015. *Cathodic protection*. Training document, Ghana: Sinopec.)

In the corrosion cell, if other conditions are the same, the smaller the polarization of the local cathode or anode is, the bigger the self-corrosion current is.

## 2.3.2 CORROSION RATE

Because of this process, electric current flows through the interconnection between the cathode and the anode. The cathodic area is protected from corrosion damage at the expense of the metal, which is consumed at the anode. The amount of metal lost is directly proportional to the current flow. Mild steel is lost at approximately 20 pounds for each ampere flowing for a year (Tong, 2015).

The degree of polarization is a measure of how the rates of the anodic and the cathodic reactions are retarded by various environmental (concentration of metal ions) and/or surface process factors. Hence, polarization measurements can thereby determine the rate of the reactions that are involved in the corrosion process—the corrosion rate. The corrosion rate can be expressed in mil per year. Corrosion rate is calculated as:

$$\text{Corrosion rate} = \frac{KW}{(\rho At)}. \tag{2.1}$$

where:

K = constant which depends on the system used,
W = weight loss after exposure to environment (mg),
$\rho$ = density of the material (g/cm$^3$),
A = area of the material exposed (square meter),
t = time of exposure (hour)

## 2.3.3 FACTORS AFFECTING CORROSION RATE

1. Potential Difference between Anode and Cathode (Galvanic Series): Interconnecting two dissimilar metals in an electrolyte will create a corrosion cell. The strength of this cell increases as the distance within the galvanic series increases.
2. Circuit resistance – Resistivity of the Electrolyte
   Circuit resistance includes the following:
   - Resistance of the anode
   - Resistance of the cathode
   - Resistance of the electrolyte
   - Resistance of the metallic path

Increasing the resistance will reduce the corrosion rate.
3. Environmental Conditions
4. Stray Currents (Table 2.1)

Example: connecting magnesium to copper will produce a corrosion cell with a potential of about 1.5 volts.

**TABLE 2.1**
**Galvanic Series**

| Metal | Volts (CSE) |
|---|---|
| Commercially pure magnesium | −1.75 V |
| Magnesium alloy | −1.60 V |
| Zinc | −1.10 V |
| Aluminum alloy | −1.05 V |
| Commercially pure aluminum | −0.80 V |
| Mild steel (clean and shiny) | −0.50 V to −0.80 V |
| Mild steel (rusted) | −0.20 V to −0.50 V |
| Cast iron (not graphitized) | −0.50 V |
| Lead | −0.50 V |
| Mild steel in concrete | −0.20 V |
| Copper, brass, bronze | −0.20 V |
| High silicon cast iron | −0.20 V |
| Carbon, graphite, coke | +0.30 V |

## 2.4 HYDROGEN-RELEASE CORROSION AND OXYGEN-CONSUMPTION CORROSION

### 2.4.1 HYDROGEN-RELEASE CORROSION

Hydrogen-released corrosion, also known as hydrogen depolarization corrosion, is corrosion in which the cathodic process is a reduction reaction of a hydrogen ion.
   Electrode reaction:

1. Necessary conditions
   - There must exist H+ ions in electrolyte solution.
   - Anode potential (EA) must be smaller than hydrogen deposition potential (EH), that is EA<EH.

Hydrogen-releasing potential refers to the potential difference between the balance potential of the hydrogen electrode and hydrogen over-potential of the cathode under a certain cathodic current density.

2. Features
   - Hydrogen-release corrosion is mainly activation polarization, and concentration polarization can be ignored if there is no deactivation diaphragm on the surface of the metal.
   - Corrosion of metal in acid solution is related to the pH value of the solution.
   - The hydrogen-release corrosion of metal in acid solution is generally macro uniform corrosion phenomenon.
3. Main factors that affect hydrogen over-potential
   - Current density.
   - Electrode material.
   - Electrolyte solution composition and temperature (Figure 2.6).

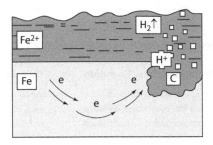

**FIGURE 2.6** Hydrogen-release corrosion. (Source: Tong, Shan. 2015. *Cathodic protection.* Training document, Ghana: Sinopec.)

### 2.4.2 OXYGEN-CONSUMPTION CORROSION

Oxygen-consumption corrosion is the corrosion in which the cathodic process is oxygen-reduction reaction.

Features: Compared to the hydrogen ion-reduction reaction, the oxygen-reduction reaction can be implemented at a higher potential so that oxygen-consumption corrosion is more universal.

1. Necessary conditions
   - There must exist oxygen in the electrolyte solution.
   - The anode potential must be lower than oxygen ionization potential, that is, EA<EO; oxygen ionization potential EO refers to the potential difference of oxygen balance potential and oxygen ionization overpotential under certain current density.
2. Features
   - If there is oxygen in the solution, oxygen-consumption corrosion will generate in the first place.
   - As oxygen diffusing is stable state, oxygen-consumption corrosion is determined by oxygen concentration polarization.
   - Dual function of oxygen: oxygen may play a role of corrosive for easy passivation of metal; it may also play a role of retardant (Figure 2.7 and Table 2.2).

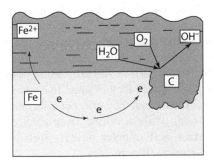

**FIGURE 2.7** Oxygen-consumption corrosion. (Source: Tong, Shan. 2015. *Cathodic protection.* Training document, Ghana: Sinopec.)

**TABLE 2.2**

**Comparison of Oxygen-Consumption and Hydrogen-Release Corrosion**

| Item | Hydrogen-Release Corrosion | Oxygen-Consumption Corrosion |
|---|---|---|
| Property of depolarizer | Three ways of mass transfer of hydrogen ion: convection, diffusion, and electromigration. The diffusion coefficient is very big. | Neutral oxygen molecule, only two ways of mass transfer: convection and diffusion. The diffusion coefficient is very small. |
| Concentration of depolarizer | Large concentration; depolarizer is hydrogen ion in acid solution and water molecule in neutral or alkaline solution. | Small concentrations; under ambient temperature and pressure, the saturation concentration of oxygen in neutral water is 10–4 mol/L, and the solubility of oxygen will decrease as temperature or salt concentration increases. |
| Cathodic reaction product | Hydrogen gas released in bubble form. | Water molecule or hydroxyl ion. Ways of leaving metal surface; convection, diffusion and electromigration. |
| Controlling types of corrosion | Cathodic control or hybrid cathodic control. Most is cathodic control, especially cathodic polarization control. | Cathodic control. |
| Influence of alloying elements or impurities | Obvious influence. | Less influence. |
| Corrosion rate | If there is no deactivation phenomenon, the corrosion rate of simplex hydrogen-released corrosion is fast because of large hydrogen ion concentration and diffusion coefficient. | If there is no deactivation phenomenon, the corrosion rate of simplex aerobic corrosion is slow because of little oxygen concentration and diffusion coefficient. |

## 2.5 CAUSES OF CORROSION

When steel is placed in an aqueous environment, electrochemical reactions take place on its surface. These consist of cathodic reaction, in which electrons are consumed, and anodic reaction, in which the metal is dissolved. The cathodic and anodic reactions are linked and their rates are related. Therefore, although the anodic reaction is the direct cause of corrosion, the rate of corrosion can be controlled by interfering with either reaction.

The corrosion which results from these electrochemical reactions may appear in various forms, e.g., general corrosion, pitting corrosion. The rate at which corrosion takes place is affected by factors such as the exact composition of the aqueous environment and the temperature. These factors need to be considered when designing corrosion protection systems, although they may not affect the type of protection used.

Other forms of corrosion, while resulting from electrochemical reactions, are more affected by design factors. Examples of these forms of corrosion are crevice corrosion and galvanic corrosion (Figure 2.8).

Corrosion is an electrochemical process occurring at the interface between metal and environment. Three conditions must be present for this to occur.

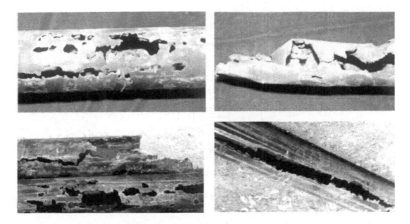

**FIGURE 2.8**  Pipeline (Equipment) corrosion pictures. (Source: Tong, Shan. 2015. *Cathodic protection*. Training document, Ghana: Sinopec.)

1. Two areas on a structure or two structures must differ in electrical potential (Figure 2.9).
2. Those areas, called **anodes** and **cathodes**, must be electrically interconnected (Figure 2.10).
3. Those areas must be exposed to a common electrolyte (soil or water) (Figures 2.11 and 2.12).

Corrosion Cell (dissimilar metals): i.e., steel pipe connected to copper ground rod. Corrosion is also caused by the formation of bacteria with an affinity for metals on the surface of the steel.

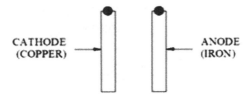

**FIGURE 2.9**  Two areas or two structures that differ in electrical potential (different amounts of stored energy).

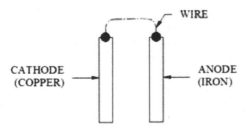

**FIGURE 2.10**  The two areas, called the "anode" and "cathode," must be electrically connected (conductive path).

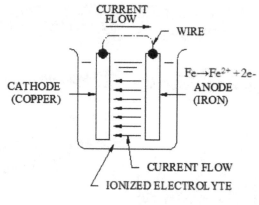

**FIGURE 2.11**   Those areas must be exposed to a common electrolyte.

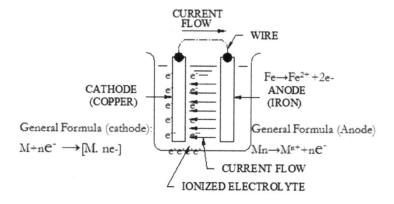

**FIGURE 2.12**   Corrosion cell (dissimilar metals).

## 2.5.1 CORROSIVE ENVIRONMENTS

The deterioration of material due to corrosion can be caused by a wide variety of environments. Metallic corrosion can occur in aqueous (i.e., water-containing) environments with or without dissolved species such as electrolytes (i.e., salts) and reactants (e.g., dissolved oxygen). However, every industry, including the food, pharmaceutical, oil, and gas, features a variety of applications encompassing a range of corrosive environments. In the food industry, the corrosion environment often involves moderately to highly concentrated chlorides on the process side, often mixed with significant concentrations of organic acids. The process environment for the pharmaceutical industry can include complex organic compounds, strong acids, and chloride solutions comparable to seawater. This book is mostly focused on the oil and gas industry, which has corrosive environments such as the sour and sweet environment from the sour reservoirs (those containing hydrogen sulfide, $H_2S$) and the sweet reservoirs (containing carbon dioxide, $CO_2$). The sour environments can result in sulfide stress cracking of susceptible materials (Taiwo, 2013).

## 2.6 CORROSION PROTECTION METHODS

There are five methods of protecting a pipeline against corrosion:

1. Material selection
2. Coatings
3. Cathodic protection
4. Chemical injection
5. Proper anticorrosion design

Coatings, which provide a physical barrier between the steel and the aqueous environment and which interfere with the flow of electrons, are covered in detail in Section 3.2 of this book. Cathodic protection, which controls the cathodic reaction, is covered in detail in Section 3.3 of this book.

For most applications, coatings and cathodic protection are used together. This ensures against coating breakdown late in the life of the system and failure of elements of the cathodic protection system at any stage. It reduces the demand on the cathodic protection system, which enables economies to be made in its design. However, where the design life is short, or where some corrosion can be tolerated, a single protection system may be appropriate.

Equipment should be designed to minimize the presence of crevices where corrosion may take place. In the case of galvanic corrosion, it is important to avoid dissimilar metals in direct contact in the same aqueous medium. Insulating flanges may be used to avoid contact. The use of coatings can minimize galvanic effects when dissimilar metals are in contact. The degree of risk from galvanic corrosion is often difficult to predict, but general guidelines on the use of a range of common materials together is given in PD 6484 (Figure 2.13).

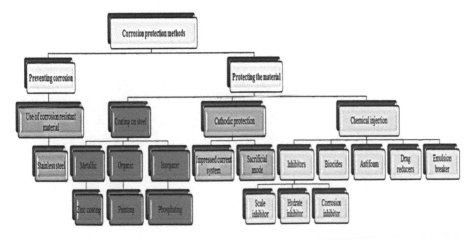

**FIGURE 2.13** Corrosion protection method.

# 3 External Corrosion Protection

External corrosion protection must be a primary consideration during design. Materials selection and coatings are the first line of defense against external corrosion. Because perfect coatings do not exist, cathodic protection must be used in conjunction with coatings. Cathodic protection involves applying a current to the pipeline through the soil from an external source and thus overriding the local anodes, rendering the entire exposed pipeline surface cathodic. Coatings function to separate the steel from the electrolyte and thus prevent corrosion.

The subsequent sections describe material selection, external coatings, and cathodic protection as applied to pipelines, subsea equipment, and structures.

## 3.1 MATERIAL SELECTION

Material selection and anticorrosion should be considered in pipeline design. Given that the surroundings of long-distance pipeline are hard to control, surface coating and electrochemistry protection are the major functions of pipeline anticorrosion.

Materials selection is critical to preventing many types of failures (CalQlata, 2011). Factors that influence materials selection are:

- Corrosion resistance in the environment
- Availability of design and test data
- Mechanical properties
- Cost
- Availability
- Maintainability
- Compatibility with other system components
- Life expectancy
- Reliability
- Appearance

Choose materials inherently resistant to corrosion in certain environments.

Material selection involves picking an engineering material, either metal alloy or nonmetal, that is inherently resistant to the particular corrosive environment and meets other criteria. Variables that will affect corrosion are established along with materials that may provide suitable resistance for those conditions. Obviously, other requirements such as cost and mechanical properties of the potential materials must be considered (Corrosionpedia, 2017).

Data needed to thoroughly define the corrosive environment include many of its chemical and physical characteristics plus application variables such as its velocity.

In addition, possible extremes caused by upset conditions need to be considered. Noncorrosion considerations include mechanical strength, type of expected loading, and possibly the compatibility of the different candidates with the required fabrication method. After these criteria and other unique ones are considered, the list of materials that can generally satisfy all requirements usually becomes short. Final selection is then made, but trade-offs often are necessary in particular cases.

Corrosion in the oil and gas industry may be prevented by upgrading to more corrosion-resistant materials (stainless steel, nickel alloys), or the corrosive environment may be treated with corrosion inhibitors or oxygen scavengers so that the traditional materials can still be used.

## 3.1.1 CONSIDERATIONS FOR MATERIAL SELECTION

The pipe material used is selected based on process conditions, the material to be conveyed, and the pipe material's ability to withstand fire and corrosion. While carbon and stainless steels are commonly used materials of construction, increasing use is being made of nonmetallic and lined or plastic process equipment. The selection of the material of construction should take into account worst-case process conditions that may occur under foreseeable upset conditions and that should be applied to all components, including valves, pipe fittings, instruments, and gauges. Both composition (e.g., chlorides, moisture) and temperature deviations can have a significant direct effect on the rate of corrosion. The operator should demonstrate that procedures are in place to ensure that potential deviations in process conditions such as fluid temperature, pressure, and composition are identified by competent persons and assessed in relation to the selection of materials of construction for pipework systems.

A wide range of plastics is available for use as materials of construction and can be used in areas such as handling inorganic salt solutions where metals are unsuitable. The use of plastic linings is widespread in equipment such as tanks, pipes, and drums. However, their use is limited to moderate temperatures, and they are generally unsuitable for use in abrasive duties. Some of the more commonly used plastics are PVC, PTFE, and polypropylene.

Special glasses can be bonded to steel, providing an impervious liner. Glass or "epoxy"-lined equipment is widely used in severely corrosive acid duties. The glass lining can be easily damaged and careful attention is required.

The following are to be considered when selecting corrosion-resistant materials for construction:

- Type of fluid to be transported (hydrocarbon, water, steam, natural gas)
- Cost-effectiveness
- Flow characteristics (laminar or turbulent flow)
- Temperature
- Pressure
- Properties of material needed
- Application of material
- Product of material (e.g., fittings, valves)

- Design
- Material versus material comparisons to find the best solution
- Industry requirements and regulations

For material to be used at room temperature, the following are the main requirements for materials selection:

- Mechanical properties of the materials: strength, hardness, toughness, resistance to fatigue and fracture
- Suitable physical properties

Ease of availability

- Easy formability
- Cost-effectiveness
- Corrosion resistance

When the materials are to be selected for high-temperature components, the requirements for material selection are:

- Mechanical properties: strength (sustaining strength at that temperature)
- Microstructural stability: change of microstructure with temperature is not acceptable
- Creep: tendency of the material to very slow deformation
- Corrosion, oxidation, and high-temperature corrosion resistance

Materials available for use in the oil and gas industry include:

- Ferrous: steel, stainless steels, super alloys
- Nonferrous: aluminum, copper, titanium, cobalt, and their alloys
- Plastics
- Ceramics
- Composites

The most commonly used material in the oil and gas industry is steel and its various types. Steel is prone to corrosion, so all steels used need to be protected by one or more of the methods of corrosion protection: coatings, cathodic protection, inhibitor, and so on (Reza et al., 2013).

It is important to select materials based on the requirements of the following international standards:

- National Association For Corrosion Engineers (NACE)
- NACE MR 175
- American Petroleum Institute (API)
- American Society for Testing and Materials (ASTM)
- American Welding Society (AWS)

- Manufacturers Standardization Society (MSS)
- American Water Works Association
- SAE International (formerly Society of Automotive Engineers)
- ASME B31

### 3.1.1.1  Material Selection Criteria for Metal Alloys

Steels are used commonly for most oil and gas applications such as piles, pipelines, piping, superstructures, support structures, storage facilities, and so on. The main material selection requirements for steel are:

- Cost-effectiveness
- Mechanical properties such as hardness, percentage elongation, yield strength, tensile strength, density, fracture toughness, creep, fatigue, brittleness
- Thermal properties such as thermal conductivity, heat capacity, thermal diffusivity
- Electromagnetic properties such as magnetic permeability/hysteresis, dielectric strength, transparency/color, electrical conductivity
- Resistance to both internal and external corrosion due to the environment (e.g., soil, groundwater, seawater, salt air)
- Ease of availability
- Fabrication
- Service performance
- Environment: operating temperature range, chemical resistance, radiation resistance, appearance

Steels may be classified as mild-, medium-, or high-carbon steel based on the percentage of carbon they contain (Reza et al., 2013).

Low-carbon steel has been identified as having a particular susceptibility to oxidation in the presence of electrolytes, water, and carbon dioxide. Therefore, the problems caused by corrosion mean that an effective solution is imperative to the prevention of accidents resulting from leakages and fractures.

### 3.1.1.2  Materials with High Corrosion Resistance

Metals that hold high corrosion-resistant properties include Incoloy 825, Inconel 625, Hastelloy C-276, Titanium, Super Duplex, Duplex, stainless steel, and 6MO.

The most commonly specified material for use in the manufacture of corrosion-resistant high-pressure valves and fittings is 316 stainless steel. It is easy to machine, cast, and forge, and is readily available as raw material in a wide range of shapes and sizes.

6MO is a 6 percent molybdenum-alloyed austenitic stainless steel. The increased molybdenum content, up to 6 percent from the 2 percent as found in 316 stainless steel, plus the increase in chromium and added nitrogen, provides 6MO with some worthwhile benefits over 316 stainless. Its corrosion resistance is increased particularly with regard to pitting and crevice corrosion. The strength of 6MO is also slightly increased above that of 316 stainless.

Monel alloy 400 is still an excellent choice where atmospheric corrosion is an issue, particularly in seawater.

Titanium: Upon exposure to air, titanium immediately forms a passive oxide coating that protects the bulk metal from further oxidation. This protective layer renders titanium immune from attack by most acids and chlorine solutions. Since titanium depends for its corrosion resistance on the presence of an oxide film, it follows that it is significantly more resistant to attack in oxidizing solutions.

Duplex and Super Duplex: These materials are a natural step up from normal stainless steels and 6MO stainless. Their structure is an amalgamation in equal parts of austenite and ferrite. As a result, these high chromium alloys have a higher tensile strength and good corrosion resistance. The super duplex alloy has excellent atmospheric corrosion resistance and finds many uses in aggressive and seawater environments. It is particularly beneficial in resisting stress corrosion cracking, which is a common problem with austenitic stainless steels. Although generally specified for use in structures and fabrication work, it is often used for pipeline components.

Incoloy 825: This alloy will give excellent corrosion resistance in a broad base of severe environments. It has good resistance to stress corrosion cracking, and the high nickel content in conjunction with the copper and molybdenum gives it outstanding resistance to reducing environments such as those containing sulfuric and phosphoric acids.

Stainless steel: Steel alloyed with at least 10 percent chromium is called stainless steel. The addition of chromium causes the formation of a stable, very thin (few nanometers) oxide layer (passivation layer) on the surface. Stainless steel, therefore, does not readily corrode or stain when in contact with water, like carbon steel does. But under some circumstances the passivation layer can break down, causing local attack such as pitting corrosion. Pitting corrosion, as the predominant form of corrosion of stainless steel, does not allow lifetime prediction as is possible with zinc coatings. The resistance of stainless steels against pitting corrosion can be roughly estimated by the PREN (pitting resistance equivalent number). The PREN is based on the chemical composition of steel, taking into account the amount of chromium, molybdenum, and nitrogen. In the literature, various equations for this calculation are given. The most common equations are the following:

*For stainless steels with Mo<3 percent*

$$PREN = \%Cr + 3.3 \times \%Mo \tag{3.1}$$

*For stainless steels with Mo≥3 percent*

$$PREN = \%Cr + 3.3 \times \%Mo + 30 \times \%N \tag{3.2}$$

There are various grades of stainless steel with different levels of stability. The most common grade is alloyed with around 18 percent Cr and 10 percent Ni. Increasing or reducing the amount of specific elements in the steel changes its corrosion properties, its mechanical properties, or some processing properties such as weldability (Hilti, 2015). If the nickel content is significantly reduced, the alloy phase will no longer be purely austenitic but will then combine austenitic and ferritic phases (duplex stainless steel).

Refer to the EN ISO 3506-1:2009 standard (mechanical properties of corrosion-resistant stainless steel fasteners, bolts, screws and studs).

## 3.2  EXTERNAL COATINGS

A coating provides a physical barrier between the steel and the aqueous environment that interferes with the flow of electrons.

This section deals with coating materials used for onshore and offshore pipelines, and gives information on the various coating systems available. Coating application, repair, use with thermal insulation coatings, under concrete weight coatings, and in conjunction with cathodic protection are discussed in this section.

### 3.2.1  STANDARDS

Numerous standards cover the application and repair of the various coating systems, and these are detailed in the sections dealing with each coating material. General guidance on the use of coatings is given in the following documents:

1. API Bull 91 Planning and Conducting Surface Preparation and Coating Operations for Oil and Natural Gas Drilling and Production Facilities in a Marine Environment
2. API RP 5L2 RP for Internal Coating of Line Pipe for Non-Corrosive Gas Transmission Service
3. API RP 5L7 RP for Un-primed Internal Fusion Bonded Epoxy Coating of Line Pipe
4. API RP 5L9 External Fusion Bounded Epoxy Coating of Line Pipe
5. API RP 652 Linings of Aboveground Petroleum Storage Tank Bottoms
6. API 1160 Managing System Integrity for Hazardous Liquid Pipelines
7. API 2217A Guidelines for Work in Inert Confined Spaces in the Petroleum Industry
8. ASTM 06.02 Paint-Products and Applications; Protective Coatings; Pipeline Coatings
9. ASTM G 8 Test Method for Cathodic Disbonding of Pipeline Coatings
10. ASTM G14 Standard Test Method for Impact Resistance of Pipeline Coatings (Falling Weight Test).
11. ASTM G17 Standard Test Method for Penetration Resistance of Pipeline Coatings (Blunt Rod).
12. DNV RP-F102 Pipeline Field Joint Coating and Field Repair of Line Pipe Coating
13. DNV RP–F106 Factory Applied External Pipeline Coatings for Corrosion Control.
14. EN 10288 Steel Tubes and Fittings for Onshore and Offshore Pipelines - External Two Layer Extruded Polyethylene Based Coatings
15. EN 10289 Steel Tubes and Fittings for Onshore and Offshore Pipelines - External Liquid Applied Epoxy and Epoxy-Modified Coatings

16. EN 10290 Steel Tubes and Fittings for Onshore and Offshore Pipelines External Liquid Applied Polyurethane and Polyurethane-Modified Coatings
17. EN 10300 Steel Tubes and Fittings for Onshore and Offshore Pipelines - Bituminous Hot Applied Materials for External Coating
18. EN 10301 Steel Tubes and Fittings for Onshore and Offshore pipelines - Internal Coating for the Reduction of Friction for Conveyance of Noncorrosive Gas
19. EN 10310 Steel Tubes and Fittings for Onshore and Offshore Pipelines - Internal and External Polyamide Powder Based Coatings
20. EN 10329 Steel Tubes and Fittings for Onshore and Offshore Pipelines - External Field Joint Coatings
21. EEMUA 194 Guidelines for Materials Selection and Corrosion Control for Subsea Oil and Gas Production Equipment
22. ISO 15741 Paints and varnishes - Friction-reduction coatings for the interior of on- and offshore steel pipelines for noncorrosive gases
23. ISO 20340 Paints and Varnishes – Performance Requirements for Protective Paint Systems for Offshore and Related Structures
24. ISO 21809 Petroleum and Natural Gas Industries – External Coatings for Buried or Submerged Pipelines Used in Pipeline Transportation Systems
25. Part 1: Polyolefin Coatings (3-Layer PE and 3-Layer PP) (In preparation)
26. Part 2: Fusion-Bonded Epoxy Coatings (Issued 2007) Part 3: Field Joint Coatings (Issued 2008)
27. Part 4: Polyethylene Coatings (2-Layer PE) (in preparation) Part 5: External Concrete Coatings (in preparation)
28. NACE 30105 Electrical Isolation/Continuity and Coating Issues for Offshore Pipeline Cathodic Protection Systems
29. NACE RP0105 Standard Recommended Practice Liquid-Epoxy Coatings for External Repair, Rehabilitation, and Weld Joints on Buried Steel Pipelines
30. NACE RP-0176 Corrosion Control of Steel, Fixed Offshore Platforms Associated with Petroleum Production
31. NACE RP0178 Fabrication Details, Surface Finish Requirements, and Proper Design Considerations for Tanks and Vessels To Be Lined for Immersion Surface
32. NACE RP0188 Discontinuity (Holiday) Testing of Protective Coatings
33. NACE RP0191 Application of Internal Plastic Coatings for Oilfield Tubular Goods and Accessories
34. NACE RP0193 External Cathodic Protection of On-Grade Carbon Steel Storage Tank Bottoms
35. NACE RP0198 Control of Corrosion Under Thermal Insulation and Fireproofing Materials - A Systems Approach
36. NACE RP0274 High-Voltage Electrical Inspection of Pipeline Coatings
37. NACE RP0303 Standard Recommended Practice Field-Applied Heat-Shrinkable Sleeves for Pipelines: Application, Performance, and Quality Control
38. NACE RP0304 Design, Installation, and Operation of Thermoplastic Liners for Oilfield Pipelines

39. NACE RP0375 Field-Applied Underground Wax Coating Systems for Underground Pipelines: Application, Performance, and Quality Control
40. NACE RP0394 Application, Performance, and Quality Control of Plant-Applied, Fusion-Bonded Epoxy External Pipe Coating
41. NACE RP0399 Plant Applied, External Coal Tar Enamel Pipe Coating Systems: Application, Performance, and Quality Control
42. NACE RP0402 Field-Applied Fusion-Bonded Epoxy (FBE) Pipe Coating Systems for Girth Weld Joints: Application, Performance, and Quality Control
43. NACE RP0602 Field-Applied Coal Tar Enamel Pipe Coating Systems: Application, Performance, and Quality Control
44. NACE RP0892 Coatings and Linings over Concrete for Chemical Immersion and Containment Service
45. NACE SP0108 Corrosion Control of Offshore Structures by Protective Coatings
46. NACE SP0169 Control of External Corrosion on Underground or Submerged Metallic Piping Systems
47. NACE SP0181 Liquid-Applied Internal Protective Coatings for Oilfield Production Equipment
48. NACE SP0185 Extruded Polyolefin Resin Coating Systems with Soft Adhesives for Underground or Submerged Pipe
49. NACE SP0490 Holiday Detection of Fusion-Bonded Epoxy External Pipeline Coatings of 250 to 760 μm (10 to 30 Mils)
50. NACE TM0102 Measurement of Protective Coating Electrical Conductance on Under- ground Pipelines
51. NACE TM0104 Offshore Platform Ballast Water Tank Coating System Evaluation
52. NACE TM0105 Test Procedures for Organic-Based Conductive Coating Anodes for Use on Concrete Structures
53. NACE TM0174 Laboratory Methods for the Evaluation of Protective Coatings and Linings Materials on Metallic Substrates in Immersion Service
54. NACE TM0185 Evaluation of Internal Plastic Coatings for Corrosion Control of Tubular Goods by Autoclave Testing
55. NACE TM0186 Holiday Detection of Internal Tubular Coatings of 250 to 760 μm (10 to 30 Mils) Dry Film Thickness
56. NACE TM0204 Exterior Protective Coatings for Seawater Immersion Service
57. NACE TM0299 Corrosion Control and Monitoring in Seawater Injection Systems
58. NACE TM0304 Offshore Platform Atmospheric and Splash Zone Maintenance Coating System Evaluation
59. NACE TM0384 Holiday Detection of Internal Tubular Coatings of Less Than 250 Micrometers (10 mils) Dry-Film Thickness
60. NACE TM0404 Offshore Platform Atmospheric and Splash Zone New Construction Coating System Evaluation
61. NORSOK M-501 Surface Preparation and Protective Coating

### 3.2.2 COATING PHILOSOPHY AND SELECTION

Coatings are applied to pipelines and subsea structures in order to insulate the steel surface from the corrosive environment, thereby preventing corrosion and subsequent failure of the structure.

Before choosing a coating material to be applied to pipelines and structures, it is important to consider the following factors:

1. The environment surrounding the pipe.
2. Pipeline operating temperature (product temperature).
3. Ease of application (yard or site): The coating should be capable of application in the factory or the field at a reasonable rate, and it should be possible to handle the pipe as soon as possible without damaging the coating.
4. Inspection methods and ease of repair of defect and pipe joints.
5. Comparative coating system costs.
6. Coating properties:
   - The coating must be resistant to the action of soil bacteria.
   - Water resistance: The coating must demonstrate negligible absorption of water and must be highly impermeable to water or water vapor transmission.
   - Resistance to disbonding: The coating must be capable of withstanding impact without cracking or disbonding.
   - The coating must be capable of withstanding stresses imposed on it by soil/pipe movement.
   - The coating must be able to form an excellent bond to the pipe steel with the use of a suitable primer where necessary. In addition, for multilayer systems, there must be good adhesion between the different coating layers.
   - The coating should show no tendency to creep under prevailing environmental conditions and must have sufficient resistance to not be displaced from the underside of large-diameter pipes.
   - The coating must have sufficient roughness to prevent slippage between the concrete coating and the anticorrosion coating.
   - The coating must be an electrical insulator of high dielectric strength and must not contain any conducting materials.
   - The coating must be sufficiently flexible to withstand deformation due to bending, hydrotesting, laying, or any expansion and contraction due to temperature change; cracks must not occur during cooling after application or curing.
   - The properties and adhesion qualities of the coating must be sufficient to withstand certain methods of installation, particularly where pipelay is by bottom tow.
   - Resistance to flow: The coating should show no tendency to flow from the pipe under the prevailing environmental conditions and must have sufficient resistance not to be displaced from the underside of large-diameter pipes.

7. Environmental and health constraints.
8. Pipeline design life.
9. Ease of field jointing.
10. Conformance to standards such as DIN 30670, NFA 49-711.

### 3.2.3  COAL TAR AND ASPHALT COATINGS

These hot applied enamel coatings have a long and successful record of accomplishment throughout the world, both onshore and offshore. However, recognition of possible health hazards during the application of asphalt enamel has contributed to reduction/withdrawal of their use. The standard BS 4147 (BS EN 10300:2005), is relevant to the materials and applications of these coatings:

- Coal tar enamels generally absorb less water than asphalts.
- Coal tar enamels adhere better to clean steel than asphaltic enamels, although this difference is not marked under normal pipeline conditions.
- Although both enamels are soluble in chemical solvents, asphalt is dissolved by petroleum products, while coal tar is only softened.
- These coatings should not be used where the pipe operating temperature is likely to exceed 60°C for normal grade coatings. However, higher temperatures can be withstood with higher grade coatings.
- Both coal tar enamel and asphalt coatings are still widely used for submarine lines under concrete weight coatings.

#### 3.2.3.1  Coal Tar Enamel

Coal tar is a by-product of coke production. The coal tar pitch is mixed with coal tar oils and inert filler such as talc or slate dust to produce coal tar enamel. The exact quantities affect properties such as hardness, flexibility, and so on.

After cleaning the pipe, a quick-drying primer is applied and allowed to dry. Following this, the coal tar is applied by flooding, and during this process layers of glass fiber mat and/or felt are drawn into the enamel to act as reinforcement.

The overall coating thickness is normally in the range of 2.5 mm–6 mm, depending on the particular specification.

#### 3.2.3.2  Asphalt Enamel

Asphalt enamels are manufactured as an end product from the refining of crude oil. It consists of two types:

1. A straight residue from distillation, which can be of the hard, high melting point.
2. The "blown" grades that are prepared by partially oxidizing the asphalt base by blowing in air.

Application methods and reinforcement are identical to that used for coal tar enamels.

### 3.2.3.3   Advantages and Disadvantages

Although the two types of enamel appear similar, they are not compatible with each other, and their performance differs markedly:

- Coal tar enamels generally absorb less water than asphalts.
- Minimum holiday susceptibility.
- Good adhesion to steel. Coal tar enamels generally adhere better to clean steel than asphaltic enamels, although this difference is not marked under normal pipeline conditions.
- These coatings should not be used where the pipe operating temperature is likely to exceed 60°C for normal grade coatings. However, higher temperatures can be withstood with higher grade coatings.
- Good resistance to cathodic disbondment.
- Although both enamels are soluble in chemical solvents, asphalt is dissolved by petroleum products, while coal tar is only softened.
- Both coal tar enamel and asphalt coatings are still widely used for submarine lines under concrete weight coatings.
- Low current requirement.
- Limited manufacturers and applicators.
- Health and air quality concerns.
- Change in allowable reinforcement.

### 3.2.3.4   Field Joints and Coating Repairs

Coating of pipeline weld joints in the field can be achieved by manual application of coal tar or asphalt, or by the application of a self-adhesive laminate tape system. Due to the inherent risks associated with the use of a high-temperature coating under field conditions, this method is rarely used onshore or offshore. Therefore, cold-applied tapes are normally used for field joints and large repairs. Small coating repairs may be carried out by "flaming over" the coating with a gas torch and scraper.

### 3.2.4   Fusion Bonded Epoxy Coatings

Fusion bonded epoxy (FBE) coatings have gained wide acceptance in the pipeline industry and can be applied on small and large-diameter pipes. Standards applicable to this type of coating are ANSI/AWWA C213, BGC/PS/CW6, ASTM A972/A972M - 00(2015), and so on.

### 3.2.4.1   Description

FBE coatings consist of thermosetting powders that are applied to a white metal blast-cleaned surface by electrostatic spray. The pipe is preheated to 230°C, and the quantity of residual heat determines the maximum coating thickness that can be achieved.

Following application, the powder melts, flows, and cures to produce thickness between 250 and 650 microns, following which the pipe is cooled by water quenching. Thickness nearer the maximum is generally required where concrete weight coating is to be applied by impingement methods.

Due to the nature of the coating, strict control of the fusion process is necessary to ensure a satisfactory coating quality (Figure 3.1).

### 3.2.4.2   Advantages and Disadvantages

- FBE coatings have been chosen in preference to coal tar enamel to achieve stability at higher temperatures and greater resistance to soil stresses. FBE coatings are suitable for use on pipes with a service temperature of up to 100°C.
- Excellent resistance to cathodic disbondment. Improvements in the performance of FBE coatings have been achieved by carrying out a precoating chromate conversion treatment of the pipe immediately after initial surface preparation. This has led to significant improvements in cathodic disbondment and hot-water immersion resistance.
- Improvements in adhesion between the concrete and FBE coating can be achieved by increasing the surface roughness.
- Although pipe steel quality has little effect on the application of enamel coatings, special consideration must be given to the pipe steel when applying epoxy powders. Due to the relative thinness of FBE coatings, it is necessary after blast cleaning to thoroughly inspect and remove all those surface irregularities which may cause defects in the finished coating.
- Although the coating is significantly harder than enamels, and is able to withstand direct impacts, angular impacts will damage the coating.
- Low current requirement.
- Lower impact and abrasion resistance.

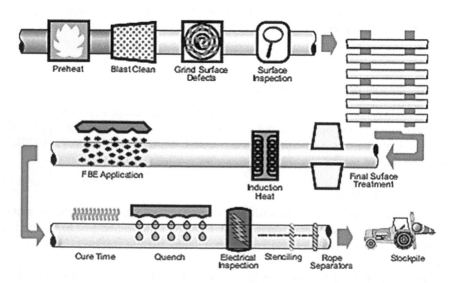

**FIGURE 3.1**   Fusion-bonded epoxy powder application schematic. (Source: Peabody's Control of Pipeline Corrosion, 2nd Edition, p. 18.)

- Excellent adhesion to steel.
- Subject to steel pipe surface imperfections. Due to the relative thinness of FBE coatings, it is necessary after blast cleaning to thoroughly inspect and remove all those surface irregularities which may cause defects in the finished coating.
- Excellent resistance to hydrocarbons.
- High application temperature.
- High moisture absorption.

### 3.2.4.3   Field Joints and Coating Repairs

Coating of pipeline weld joints may be carried out in the field using portable blast cleaning, induction heating, and powder spray equipment.

Alternatively, urethane mastic or laminate tape wrap may also be used.

### 3.2.5   POLYETHYLENE COATINGS

Polyethylene coatings are a relatively recent innovation, and consequently less is known about their long-term performance than FBE coatings.

### 3.2.5.1   Description

Polyethylene coatings may be applied by one of the following three processes:

1. Circular or ring-type head extrusion
2. Powder sintering
3. Side extrusion and wrapping

In addition, the extruded polyethylene may be applied in conjunction with primer or adhesive systems.

Polyethylene coating systems favor the use of a high-density polyethylene with either butyl rubber or hot applied mastic adhesives.

Improved adhesion and resistance to cathodic disbondment can be achieved by priming the pipe surface first with an epoxy-based layer on top of which the adhesive layer and polyethylene coating is applied. This three-layer system is now readily available from various coating plants.

The DIN 30670 standard is applicable to polyethylene coatings. It notes that although a thickness of 1 mm is sufficient for corrosion protection, the mechanical load-bearing capacity of the coating may be improved by using a thickness of between 1.8 and 3 mm, depending on the pipe diameter.

Sintered polyethylene coatings have recently become available and are produced by pouring polyethylene granules onto preheated steel pipe at 450°C.

Polyethylene coatings should not be used where the pipe service temperature is likely to exceed 65°C. They are resistant to damage during handling and laying. Where concrete is to be used over a polyethylene, consideration should be given to increase surface roughness. Coating disbonding may occur due to elongation of the pipe caused by a rise in temperature (Figure 3.2).

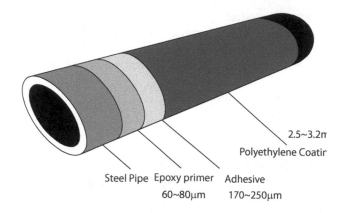

FIGURE 3.2    Schematic Diagram of 3LPE. (Source: Tong, Shan. 2015. *Cathodic protection.* Training document, Ghana: Sinopec.)

### 3.2.5.2    Advantages and Disadvantages

- Polyethylene coatings provide a tough and rugged corrosion protection system for pipelines, with excellent resistance to damage during handling and laying. Although the material can withstand impacts during handling, they do affect the adhesion of the coating, particularly with a hard copolymer adhesive.
- Polyethylene coating should not be used where the pipe service temperature is likely to exceed 65°C.
- Where concrete is to be used over polyethylene, consideration should be given to increase the surface roughness so that a better key between the concrete and the polyethylene coating is achieved.
- Problems have also been experienced with the high-density polyethylene/rubber adhesive system, where poor quality control during manufacture has led to variable application of the adhesive. This particular system has shown poorer resistance to above-ground exposure with embrittlement of the polyethylene and aging of the adhesive. In addition, cases have been reported of the coating disbonding due to elongation of the pipe caused by a rise in temperature.
- High-tolerance to various kinds of corrosions in different environments.
- Favorable price/performance ratio.
- Favorable insulation performance; immune to long-term dry and immersion conditions.
- High strength, not easily damaged by backfill gravels of which the diameter is less than 25 mm.
- Anti-disbonding.
- Durable service life: 50 years or more under 60°C.

### 3.2.5.3    Field Joint and Coating Repair

Welded pipe joints are coated using either heat-shrink polyethylene sleeves or cold-applied self-adhesive laminate tapes. Field joints are normally prepared by wire brushing, although shot blasting is sometimes preferred for sleeves.

Heat-shrink sleeves require careful application to ensure a satisfactory bond is obtained.

Small coating repairs either are made with hot-melt polyethylene sticks, polyethylene sheet patches, or tape wrap.

## 3.2.6 TAPE WRAP

Tape wrap coatings are normally used for field joints, coating repairs and for coating short lengths of pipework in the field. They are not usually applied at the factory due to their inherent susceptibility to damage during transportation and handling. They are normally applied in the field by hand or machine direct from a roll of tape. These tapes are produced in both temperate and tropical grades, and heavy duty versions are available for use at field joints under hot mastic asphalts or concrete weight coating.

The AWWA C209 standard is applicable to tape wrap coatings.

### 3.2.6.1 Self-Adhesive Bituminous Laminate Tapes

These tapes normally comprise a bituminous adhesive compound layer applied to a PVC backing of varying thickness between 0.08 to 0.75 mm, giving a total tape thickness of approximately 2 mm.

Some tapes are also manufactured with a fabric reinforcement within the bituminous layer to improve impact resistance.

To achieve the required coating thickness, the tape is applied with either a 25 mm or 55 percent (of tape width) overlap using a purpose-made machine or by hand.

## 3.2.7 EPOXY AND URETHANE LIQUID COATINGS

Although the previous sections have summarized the various coating systems generally used for the corrosion protection of pipelines, a number of other paint systems exist which may be used for specific applications where a high degree of chemical or abrasion resistance is required. These paint systems are based on epoxy and urethane resins and may be used to coat short lengths of subsea pipelines and protection structures after fabrication.

### 3.2.7.1 Description

The paint systems normally considered for use on pipelines are the two-pack high build systems. Preparation of the pipe surface by abrasive blasting is generally required for these coatings in order to achieve a satisfactory bond to the steel. Following this, the paint is normally applied to the required thickness in one coat using airless spray equipment.

Normally, depending on the required service life and conditions, between 1–5 mm would be applied.

Both polyurethanes and epoxies can be used with a coal tar in order to reduce the material cost.

### 3.2.7.2   Advantages and Disadvantages

- Due to their resistance to abrasion, they are useful for thrust-bored pipe installations, and their chemical resistance makes them particularly useful where pipework may be exposed to petroleum or chemical products.
- These coatings are particularly useful for short lengths of tie-in pipework and subsea structures.
- When compared to other coating systems, their main disadvantage is one of cost. In addition, application is necessary under strictly controlled conditions, and special spraying equipment for hot applied polyurethanes is necessary. This is because polyurethanes typically contain iso-cyanates that are linked directly to respiratory problems. In some locations, their use may not be permitted for health reasons.
- Coal tar epoxy and urethane coatings are not recommended for use on pipelines where the service temperature is likely to exceed 80°C.

### 3.2.7.3   Field Joint and Coating Repairs

Coating damage can be repaired by using small quantities of the paint, and field joints can be coated using versions which cure more quickly and which can be hand-painted/troweled on to the pipe surface. As with the pipe coating, field joints require abrasive blasting.

### 3.2.8   Coal Tar Epoxy Coatings

A coal tar epoxy is a black surface protection polymer used on surfaces subjected to extremely corrosive environments. It is a blend of various epoxy resins and coal tar. Coal tar epoxy is made by the conversion of polyamide epoxy with a pitch of refined coal tar. It is mostly used on metal substrates and concrete in offshore, petroleum, and industrial environments. It is commonly used to make high solids coatings or paints to provide moisture protection for underground systems like pipelines, water treatment facilities, clarifiers, and tanks; it is further used in the sewage industry and for prevention of microorganisms. There are different types of paints: two-component paint, three-component paint, and so on. The mixture is used in two-component paints.

The fluctuation of temperatures can make the product crystallize. It is stable at room temperature or at least 5°F above dew point. The environmental conditions affect the drying time of the product (Corrosionpedia, 2017).

### 3.2.8.1   Advantages and Disadvantages

- It forms paints or coatings that are smooth in brush, roller, and spray application.
- It bonds well with oily surfaces and is hence preferable in garages.
- It forms a good moisture sealing for paints or coatings.
- It produces paints or coatings with abrasion, thermal shock, impact, and chemical resistance and is suitable for sustained immersions in saline or freshwater.
- Provides maximum corrosion protection.
- Provides protection against soil stress.

- Provides a semi-gloss with a matte surface finish.
- It provides self-priming and good adhesion in paints or coatings.
- Coal tar epoxy and urethane coatings are not recommended for use on pipelines where the service temperature is likely to exceed 80°C.

### 3.2.9 MILL-APPLIED TAPE COATING SYSTEMS

The mill-applied tape systems consist of a primer, a corrosion-preventative inner layer of tape, and one or two outer layers for mechanical protection. Concern regarding shielding of Cathodic Protection (CP) on a disbonded coating has led to development of fused multilayer tape systems and of a backing that will not shield CP (Sloan, 2001).

#### 3.2.9.1 Advantages and Disadvantages

- Good adhesion to steel
- Minimum holiday susceptibility
- Ease of application
- Low energy required for application
- Handling restrictions—shipping and installation
- UV and thermal blistering— storage potential
- Shielding CP from soil
- Stress disbondment

### 3.2.10 EXTRUDED POLYOLEFIN SYSTEMS

There are two extrusion types available:

1. Crosshead extrusion
2. Side extrusion

Crosshead extrusion consists of an adhesive and an extruded polyolefin sheath. This system is limited to pipe diameters 1/2 in. through 36 in. (13 mm through 900 mm).

Side extrusion consists of an extruded adhesive and an external polyolefin sheath. This system is limited to pipe diameters 2 in. through 146 in. (50 mm through 3650 mm).

Extruded polyolefin coatings have performed at higher temperatures. Recent improvements in the adhesive yield better adhesion, and selection of polyethylene has increased stress crack resistance (Sloan, 2001, AWWA C215 1999).

Available with polypropylene for use at higher temperatures (up to 190°F [88°C]), these systems have been used in Europe since the mid-1960s, along with the side extrusion method for larger diameters, through 60 in (152.4 cm).

#### 3.2.10.1 Crosshead-Extruded Polyolefin with Asphalt/Butyl Adhesive

*3.2.10.1.1 Advantages and Disadvantages*

- Minimum holiday susceptibility
- Low current requirements

- Ease of application
- Nonpolluting
- Low energy required for application
- Minimum adhesion to steel
- Limited storage (except with carbon black)
- Tendency for tear to propagate along pipe length

### 3.2.10.2   Dual-Side-Extruded Polyolefin with Butyl Adhesive

*3.2.10.2.1   Advantages and Disadvantages*
- Minimum holiday susceptibility
- Low current requirements
- Excellent resistance to cathodic disbondment
- Good adhesion to steel
- Ease of application
- Nonpolluting
- Low energy required for application
- Difficult to remove coating
- Limited applicators

### 3.2.11   MULTILAYER EPOXY/EXTRUDED POLYOLEFIN SYSTEMS

First introduced in Europe in the mid-1960s as a hard adhesive under polyethylene, followed by the addition of an epoxy primer (FBE or liquid), multilayer epoxy/polyolefin systems are the most-used pipe coating systems in Europe. These systems are now available throughout the world (Sloan, 2001) (Figure 3.3).

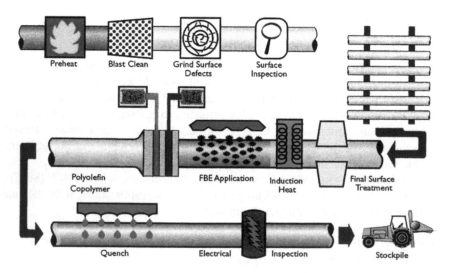

**FIGURE 3.3**   3–Layer copolymer coating application schematic. (Source: Peabody's Control of Pipeline Corrosion, 2nd Edition, p. 28.)

### 3.2.11.1  Advantages and Disadvantages
- Lowest current requirements
- Limited applicators
- Highest resistance to cathodic disbondment
- Exacting application parameters
- Excellent adhesion to steel
- Higher initial cost
- Excellent resistance to hydrocarbons
- Possible shielding of CP current
- High impact and abrasion resistance

## 3.2.12  ELASTOMER COATINGS

A number of elastomer materials are used for coatings. They include:

1. Natural rubber
2. Polychloroprene (neoprene)
3. EPDM (Ethylene Propylene Diene Monomer)

Elastomer coatings are generally more expensive than other coating systems described in this book, so their use is restricted to specific applications where other coatings are inadequate. These include the following:

1. Splash zone coating of risers, where there is high potential for corrosion, no effective cathodic protection, and a risk of mechanical damage.
2. Pipelines with high fluid temperatures—for example, EPDM is suitable for operating temperatures up to 130°C, and polychloroprene coatings are suitable for operating temperatures up to 95°C.
3. Applications where corrosion protection and thermal insulation are required, in which the elastomer is normally part of a formulation that includes other materials such as PVC foam, polyurethane foam, or glass spheres.
4. Polychloroprene (PCP) can also be internally applied to spools and can be used as a lining for riser clamps or as an external anti-fouling coating by the application of a 3mm-thick outer layer of PCP containing copper nickel granules (this system is called cupoprene).
5. Elastomer coating can be used for coating straight lengths of pipe, field joints, and custom parts such as bends and fabricated spools.

### 3.2.12.1  Field Joint and Coating Repairs
Various field joint systems are available for elastomer coatings. They include the use of uncured elastomer that is cured in situ by electrical heating and the use of formulations that involve PVC foam or mastic materials.

Most elastomer coatings are proprietary systems, and new systems are developed frequently. It is therefore necessary to consult manufacturers about detailed specifications for systems currently available.

### 3.2.13   High-Temperature Coatings

This section deals with coating materials used for onshore and offshore pipelines operating at elevated temperature and gives information on various coating systems available for thermal insulation. Other coatings for equipment and structures at elevated temperatures are also covered.

A key limiting factor for pipelines transporting hot products is the external corrosion coating. A small number of external pipeline coatings are effective at operating temperatures above 66°C (150°F). As the oil and gas industry strives to cut costs and improve profits, high-temperature pipeline coatings become more attractive.

High-temperature coatings are mostly used in the aerospace, manufacturing, military, petrochemical, and power industries for piping, fireproofing, jet engines, offshore rigs, original equipment, and various type of plants/facilities that employ high-temperature processes. One of the largest users of industrial high-temperature coatings is processing facilities such as power plants, petrochemical plants, and refineries.

Coal tar related coatings could creep to the bottom of the pipe or become brittle and crack. When this occurs, the pipeline operator must compensate with more and more cathodic protection. With extruded polyolefin, shrink sleeves and tapes, high temperatures can cause the adhesive to creep to the bottom of the pipe. This leaves a shell of polyolefin that will shield the cathodic protection. If water penetrates this shell, significant corrosion can occur. At elevated temperatures, Fusion Bonded Epoxy (FBE) can absorb more water than normal, and if underfilm contaminants are present, blisters can develop, causing the FBE to disbond. Once again, more cathodic protection has to be used to overcome the effects of the disbondment. For these reasons, the operating temperature of the pipeline must be a major factor in the selection of any external pipeline coating (Norsworthy, 1996). To prevent the water penetration in the FBE at higher temperatures, several over coats have been tried. The most successful has been the use of a chemically modified polypropylene (CMPP).

High-temperature coatings for onshore or subsea pipelines that can be used at elevated service temperatures include polypropylene coating, polyurethane elastomer, and foam materials such as syntactic foam. High-temperature coatings for process plants, structures, and equipment include epoxy novolac coatings, silicone coatings, multi-polymeric coatings, epoxy phenolic coatings, and modified silicone coatings.

#### 3.2.13.1   Standards

Although numerous standards exist which cover the properties of the raw materials for high- temperature and thermal insulation coating systems, no one standard covers their application and repair, with the exception of medium-density polyethylene (MDPE) and polypropylene (PP), which are covered by the DIN 30670 and NFA 49-711 standards, respectively. All coatings detailed are suitable for use at temperatures in excess of 70°C without a concrete weight coating. ASTM D2485, "Standard Test Method for Evaluating Coatings for High-Temperature Service," is used to determine a coating's resistance to elevated temperature and a corrosive environment. Resistance to cathodic disbondment is evaluated in accordance with ASTM G42, "Standard Test Method for Cathodic Disbonding of Pipeline Coatings Subjected to Elevated Temperatures."

### 3.2.13.2 Coating Philosophy and Selection

High-temperature and thermal insulation coatings are applied to pipelines to isolate the steel surface from the corrosive environment and provide the pipeline with thermal insulation. Before a choice of coating can be made, it is necessary to consider the following factors:

- The environment surrounding the pipe
- The pipeline operating temperature
- Ease of application (yard or site)
- Inspection methods and ease of repair of defect and pipe joints
- Comparative coating system costs
- Coating properties
- Environment and health constraints
- Degree of thermal insulation required
- Pipeline design life
- Ease of field jointing

Following consideration of the items listed above, selection of a suitable coating material for the envisaged service conditions may now be carried out. As there are limitations to each type of coating material, an ideal coating system would possess the following characteristics:

- Ease of application: The coating should be capable of application in the factory at a reasonable rate, and it should be possible to handle the pipe as soon as possible without damaging the coating.
- The coating must be able to form an excellent bond to the pipe steel with the use of a suitable primer where necessary. In addition, for multilayer systems, there must be good adhesion between the different coating layers.
- The coating must be capable of withstanding stresses imposed on it by soil/pipe movement.
- The coating must be capable of withstanding impact without cracking or disbonding. Should the outer layer be breached, then the underlying layers should resist water penetration/absorption.
- The coating must be sufficiently flexible to withstand any deformation due to the forces associated with S-lay, reel barge installation, and any expansion and contraction due to temperature change.
- The coating must be resistant to the action of soil bacteria.
- The coating must demonstrate negligible absorption of water and must be highly impermeable to water or water vapor transmission.
- The coating should show no tendency to creep under the prevailing environmental conditions and must have sufficient resistance to not be displaced from the underside of large-diameter pipes.
- The coating must have sufficient inherent roughness, or be able to have its surface roughness increased without damaging the coating, to prevent slippage between the coating and a concrete weight coating (when applied).

### 3.2.13.3   Polypropylene Coatings

With the exploitation of increasingly hotter well fluid temperatures both onshore and offshore, polypropylene (PP) as a coating material has been increasingly used. In order for the material to maintain its stability at elevated temperatures (>100°C), the polypropylene coating is normally chemically modified.

Although the DIN 30670 standard does not deal specifically with polypropylene coatings, it is nevertheless frequently referred to for their application and testing. The recently published French standard NF A49-711 and ISO 9001:2008 deal specifically with the application and testing of two- and three-layer polypropylene coatings (Chalke and Hooper, 1994).

#### 3.2.13.3.1   Description

Polypropylene coatings may be applied to linepipe by one of two processes:

- Side extrusion and wrapping
- Circular or ring-type head extrusion

The polypropylene is normally supplied in conjunction with an FBE primer either with or without an intermediate adhesive system, depending on whether a two- or three-layer system is being used.

United States pipe coaters favor the two-layer polypropylene system employing a 400–500micron thickness of FBE and overlaying this with a similar thickness of chemically modified polypropylene.

European pipe coaters, however, favor the use of a three-layer polypropylene system comprising a primer layer of FBE some 70–150 m thick, followed by a copolymer adhesive some 230–400 m thick and the chemically modified polypropylene, which can range in thickness from 1 to 3 mm (Chalke and Hooper, 1994).

Current testing has indicated that a multilayer coating consisting of FBE 500–750 microns (20–30 mils), topped with 625–1275 microns (25–50 mils) of chemically modified polypropylene (CMPP), will perform well at internal operating temperatures of up to 150°C (300°F) with cathodic protection applied. This system could be applied as a two- or three-layer system. The three-layer system may consist of a thick FBE, with 125–250 microns (5–10 mils) of CMPP for a "tie" layer and a thick layer of unmodified polypropylene as a top coat (Norsworthy, 1996).

#### 3.2.13.3.2   Advantages and Disadvantages

- Polypropylene coatings provide a tough and rugged corrosion protection system for pipelines, with excellent resistance to damage during handling and laying. This can be an important advantage over "softer" pipe coatings such as fusion-bonded epoxy (FBE) coating and coal tar enamel (CTE).
- Although a three-layer polypropylene coating may be able to withstand an impact test of 40J without obvious external damage, adhesion between the underlying layers may be affected.

- As with polyethylene, where a concrete weight coating is to be applied over the polypropylene, the surface roughness should be increased to provide a key. This may be easily achieved by sprinkling the surface of the coating with granules of polypropylene immediately after the extruder.
- One of the main advantages of polypropylene is its stability at elevated temperatures, and for this reason it has been used on projects (both onshore and offshore) having a design temperature of up to 120°C and where there is no requirement for thermal insulation.
- As with FBE coatings, pre-treatment of the steel surface using a chromate wash is recommended in order to improve the adhesion of the FBE primer layer for the three-layer polypropylene system. In addition, pipe steel quality is of considerable importance with any raised slivers or other imperfections having to be ground flat after abrasive blasting of the pipe surface.

### 3.2.13.3.3   Field Joints and Coating Repairs

The introduction of polypropylene (PP) as a coating material for linepipe has preceded the development of a satisfactory field joint coating system, and this is particularly true for the three-layer polypropylene system.

Conventional shrink sleeves, while able to bond to the pipe steel and withstand the elevated temperatures, do not form a bond with the polypropylene and consequently must be considered unsuitable. Although a polypropylene-based shrink sleeve is being developed, indications are that effective bonding between the sleeve and coating at the overlap area is not achieved either.

Cold-applied bituminous tape wrap is considered to be unsuitable also, due to the material's inability to resist elevated temperatures.

To date, the most effective joint system is considered to be a two-layer (FBE/PP) type. Although thinner than the pipe coating (approximately 1.8 mm compared to 3 mm), it does have the advantage of being able to resist temperatures of up to 120°C. This could be accomplished by applying a two-part epoxy that is partially heat cured before chemically modified polypropylene (CMPP) powder is applied and melted with a hot, nonstick surface.

A three-layer polypropylene (PP) system has been developed (FBE/adhesive/PP) which will offer a field joint of similar quality to the linepipe coating on some projects which used three-layer PP coated pipe.

The Three-Layer Polypropylene Systems (3LPP) is a multilayer coating composed of three functional components (FBE/adhesive/PP). This anticorrosion system consists of a high-performance fusion-bonded epoxy (FBE) followed by a copolymer adhesive and an outer layer of polypropylene (PP), which provides the toughest, most durable pipe coating solution available. The 3LPP System provides excellent pipeline protection for small- and large-diameter pipelines with high operating temperatures. The FBE component of the 3LPP System provides excellent adhesion to steel, providing superior long-term corrosion resistance and protection of pipelines operating at high temperatures. The superior adhesion properties of the FBE also results in excellent resistance to cathodic disbondment, reducing the total cost of cathodic protection during the operation of the pipeline.

### 3.2.13.4    Polyurethane Elastomer

Polyurethane (PU) elastomers are high-performance solid materials formulated to provide a variety of products with a wide range of physical properties. They are distinctive with respect to their high load-bearing capacity, toughness, and abrasion resistance, and they typically have a density of 1100 $kg/m^3$.

#### 3.2.13.4.1    Advantages and Disadvantages

*   PU elastomers have high natural resistance to fatigue, hydrolysis, and ultra-violet degradation. Combined with excellent resilience, toughness, and abrasion resistance, this makes polyurethane (PU) elastomer material an excellent choice for surface coatings in a marine environment.
*   Polyurethane elastomers have a maximum service temperature limit of 115°C, and there is no limit on the water depth at which it can be used due to its elastomeric nature. The material can be applied directly to line pipe as an anticorrosion coating if required. The material has a thermal conductivity (K) value of 0.22 $W/m^2K$ and as a consequence relatively high thicknesses are required to obtain low overall heat transfer (U) values, which may make the system excessively expensive or impractical. Due to this, polyurethane elastomer coatings for thermal insulation service normally employ a foam material. Also, in terms of material costs, solid PU is slightly less than average when compared to other insulating materials. Polyurethane elastomer coatings may be used in conjunction with a concrete weight coating and can be installed by either reel or conventional laybarge.

#### 3.2.13.4.2    Field Joint Coating and Repairs

The use of polyurethane elastomer coatings for linepipe does pose some problems at the field joint area. Although castable grades of polyurethane elastomers are available for site use which are able to withstand a service temperature of 125°C, extreme care has to be taken with surface preparation of the field/pipe coating interface if satisfactory adhesion is to be achieved. This also applies to the use of shrink sleeves for field joints.

For thermally insulated pipework, pre-formed half-shells made from a foam material can be used to achieve a low U-value at the field joint. Due to its roughness, repairs to the coating are considered extremely unlikely.

### 3.2.13.5    Foam Materials

Numerous foam materials are available for use on pipelines requiring thermal insulation depending upon service temperature, water depth, and required U-value, among other things. Foams may be either open or closed cell, although closed-cell materials are preferred for subsea use due to their ability to resist significant water penetration should the outer layer become damaged. The foams most commonly used for pipeline insulation are polyurethane (PU) and polyvinyl chloride (PVC).

Both of these materials are created synthetically by a process in which a gas (called the blowing agent) is generated within a confined, heated fluid which expands to form a foam. Blowing agents such as carbon dioxide or Freon-11 have been used to replace air in the cell structure of the material in order to decrease thermal conductivity of the material.

Polyethylene foams are flexible, resilient closed-cell materials. They are chemically inert, and fire-retardant grades are available. The foam is available in a range of densities from 25 to 17.5 kg/m$^3$. The material is manufactured in roll or sheet form and therefore has to be thermo-laminated to produce greater thicknesses (Chalke and Hooper, 1994).

Glass foam is a 100 percent closed-cell rigid foam material formed by blowing glass with $H_2S$ to produce a foam with very small cell size. Because it is made from an inorganic material, it is impervious to moisture and noncombustible.

### 3.2.13.5.1   Advantages and Disadvantages

- PU and PVC foams are each able to withstand a maximum service temperature of 100°C, although both materials are liable to the effects of compressive creep. In addition, it should be noted that the K value of these materials increases with increasing temperature.
- PVC foams vary in density from 200 to 450 kg/m$^3$, and PU foams have a density of 160 kg/m$^3$. PU foam has a maximum water depth limit of 60 m, whereas PVC foam is able to withstand immersion to a depth of 200 m with no ill effects. However, both materials require covering with a solid elastomer to prevent water ingress.
- When used for thermal insulation of pipelines, both PU and PVC foams may be applied directly to the steel surface of the pipe and covered with a solid water-impermeable elastomer. However, should the outer layer be breached due to impact damage during installation or trawl boards, it is possible for corrosion to occur, as foams do not provide any degree of corrosion protection. In order to overcome this, an anticorrosion coating of either FBE, EPDM, neoprene, PP, or PU is normally applied to the linepipe first, depending upon the operating temperature of the pipeline.
- PE foam, because it is produced in roll or sheet form and has to be thermo-laminated to obtain greater thickness, has not been used for the thermal insulation of pipelines to date.
- Glass foam, on the other hand, has been used on numerous occasions for the production of half-shells for the field joints of thermally insulated pipelines. The complex manufacturing procedure does not lend itself to foam-in-place application.

### 3.2.13.5.2   Field Joints and Repairs

Depending upon the type of anticorrosion coating applied to the linepipe, it may be necessary to apply a similar or compatible system to the field joint area. For the filling of the field joint area, two methods are essentially available. One involves placing a removable mold around the joint and filling the field joint with a liquid polyurethane, which is then allowed to cure. Should a higher degree of thermal insulation be required, the use of pre-formed foam half-shells may be considered. Conventional practice is then to fill the half-shell gaps with an injected PU elastomer. Additionally, security for the half-shells may be achieved by covering the finished field joint with a heat-shrink sleeve.

As with other foam systems, repairs to the coating are considered unlikely because the foam is normally overcoated with a solid elastomer (PU or EPDM). Repairs to the coating system may be feasible but would depend largely on the severity of the damage. Therefore, either any repairs should be performed as a field joint, or the pipe joint should be re-coated.

### 3.2.13.6  Syntactic Foams

Syntactic foams are rigid materials originally developed for deep-water buoyancy. They are manufactured by welding glass microspheres with rigid resin systems. Glass microspheres are used in preference to either polymer or ceramic beads because they do not creep under the combination of pressure and temperature, and they offer a high insulation factor.

Syntactic polyurethane and polypropylene foams have maximum operating temperatures of 115°C and 120°C, densities of 1100 kg/m$^3$ and 900 kg/m$^3$, and K values of 0.22 w/m$^2$K and 0.17 W/m$^2$K, respectively (Chalke and Hooper, 1994).

#### 3.2.13.6.1  Advantages and Disadvantages

- Due to their nature, there is no restriction on the water depth at which they can be used, and both may be used on pipelines which are to be installed using conventional or reel barge techniques.
- Compared to other available foam insulation materials, the cost of syntactic PU and PP is average. The application of a separate anticorrosion coating to the surface of the linepipe may not be necessary with syntactic PU because it is an effective anticorrosion material by itself. Although syntactic PP should be capable of being similarly applied, it is currently applied over the top of an FBE coating.

#### 3.2.13.6.2  Field Joints and Coating Repairs

As with the conventional foam insulation systems, field joints and repairs to the coating may be achieved as described in Section 3.2.13.5.2.

### 3.2.13.7  Epoxy Phenolic Coatings

Epoxy phenolic coatings are classified as either ambient cure, in which the phenolic and epoxy resins chemically react at room temperature, or heat cure, where the coating is exposed to temperatures of 350-400°F to accelerate the cure or activate a catalyst or curing agent in the coating. Epoxy phenolic coatings provide chemical, solvent, and temperature resistance, and are commonly used for immersion service, tank linings, and high-temperature oil and brine immersion service, for process plants, evaporators, and so on that contain boiling water. Other appropriate applications of epoxy phenolic coatings are when severe chemical resistance is necessary but a high degree of flexibility is not (O'Malley, 2018).

### 3.2.13.7.1   Advantages and Disadvantages

- Excellent adhesion properties.
- Temperature resistance of up to 400°F. Excellent resistance to boiling water, acids, salt solutions, hydrogen sulfide, and various petroleum products.
- Resistance to solvents, chemicals, and abrasion.
- In a fully cured form, they are odorless, tasteless, and nontoxic.
- Disadvantages include decreased weatherability and flexibility, relatively slow air-curing time, and often the necessity of heat-curing at relatively high temperatures.

## 3.2.13.8   Epoxy Novolac Coatings

Epoxy novolac coatings exhibit improved heat resistance because of the presence of aromaticity in their molecular structure, coupled with more cross-linking compared to other epoxies. Novolac epoxies are typically heat resistant up to 350–360°F.

In general, novolac epoxies are known for having greater resistance to oxidizing and nonoxidizing acids, and to aliphatic and aromatic solvents, compared to other epoxies. These qualities make novolac epoxies an option for applications such as tank linings in contact with high-temperature acidic crude oil. Epoxy novolac coating is ideal for harsh chemical and solvent-resistant applications. It is used in secondary containment, solvent storage, pump pads, trenches, and other high-exposure areas (O'Malley, 2018).

## 3.2.13.9   Silicone Coatings

Silicone coatings contain resins that are either pure or hybrid polymers and consist of organic pendant groups attached to an inorganic backbone of alternating silicon and oxygen atoms. The polymer structure provides thermal stability and oxidation resistance. Silicones are essentially transparent to ultraviolet radiation from sunlight (O'Malley, 2018).

High-temperature 100-percent-silicone coatings are single component and cure by heat-induced polymerization. These thin-film paints dry by solvent evaporation to achieve sufficient mechanical strength for handling and transport. However, total cure is achieved only after exposure to temperatures in the 350–400°F range. Curing can be achieved as the equipment is returned to its operating temperature. Pure silicone coatings are used on exhaust stacks, boilers, and other exterior steel surfaces at temperatures ranging from 400°F to 1200°F.

## 3.2.13.10   Modified Silicone Coatings

These high-temperature coatings have lower resistance to elevated temperatures than 100-percent-silicone coatings. Silicone acrylics are single-package air-dry paints that have color and gloss retention to temperatures in the 350–400°F range. Similarly, silicone alkyds are single-package air-dry paints with similar color and gloss-retention properties. However, the dry heat resistance of silicone alkyds is limited to about 225°F. Although most high-temperature silicones require ambient

**TABLE 3.1**
**Coating Summary Table**

| Coating Material | Application | | Maximum Temperature (°C) | Maximum Water Depth (m) | Cost (a) |
| | Onshore | Offshore | | | |
|---|---|---|---|---|---|
| Polypropylene | √ | √ | 120 | N/A | 2 |
| Polyurethane | √ | √ | 115 | N/A | 4 |
| EPDM (Ethylene Propylene Diene Monomer) | √ | √ | 130 | N/A | 3 |
| Polychloroprene | √ | √ | 95 | N/A | 3 |
| PU foam | √ | √ | 100 | 60 | 2 |
| PVC foam | √ | √ | 100 | 200 | 10 |
| Syntactic | √ | √ | 115 | 300 | 5 |
| Syntactic PP | √ | √ | 120 | N/A | 5 |

*Source*:   P. Chalke, J. Hooper. 1994. *Addendum to corrosion protection guidelines.* JPK corrosion protection guidelines, United Kingdom: JP Kenny.

a  1 = cheapest, 10 = most expensive

---

temperatures for application, special formulations are available that can be applied to steel up to 400°F.

### 3.2.13.11   Multi-Polymeric Matrix Coatings

Multi-polymeric matrix coatings are either single- or multi-component inert and inorganic, and are composed of resin combinations. Usually, multi-polymeric coatings contain aluminum and micaceous iron oxide flake, or titanium. Results from manufacturer studies have revealed anticorrosive performance with single-coat applications (150–200 microns [6–8 mils]) between ambient and 400°C (752°F) in both atmospheric exposure and under insulation tests (O'Malley, 2018) (Table 3.1).

### 3.2.14   OTHER COATINGS

Other coatings to be considered for onshore and offshore pipelines include the use of splash zone coatings (e.g., Monel, anti-fouling paint, Nano coating, etc.).

### 3.2.14.1   Concrete Weight Coatings

Concrete weight coating (CWC) is a plant-applied coating used to provide negative buoyancy for offshore pipelines or for river or road-crossing applications. Concrete weight coating is used when stability of the pipeline on the seabed is an issue. The concrete weight coating also provides mechanical protection against dropped objects and impact by trawl boards. Concrete weight coatings cannot be used on pipelines laid by the reel method. The concrete weight coating is applied to the coated pipe by any of the following methods:

- Impingement
- Extrusion
- Slip forming

The minimum thickness for a concrete coating applied by impingement is about 40 mm, and the minimum thickness applicable by the extrusion process is 35 mm. The two common densities of concrete that are used are 140 lbs/cu. ft and 190 lbs/cu. ft. Higher density is obtained by adding iron ore to the concrete mix. Recently, higher-density iron ore has been used to obtain concrete density ranging from 275 to 300 lbs/cu. ft for the Ormen Lange pipeline in the North Sea. An intercoat may be needed over thin coatings such as FBE to protect thin epoxy coating from damage by the high-velocity impinged concrete or the concrete extrusion process (Tian et al., 2005).

## 3.3  CATHODIC PROTECTION

Cathodic protection introduces the cathodic current to the protected metal structure to eliminate the potential difference caused by the component so that the corrosive current can be reduced to zero, avoiding the electrochemical corrosion.

When a metal is placed in an aqueous environment, electrochemical reactions take place on its surface. There is a cathodic reaction, in which electrons are consumed and a simultaneous anodic reaction, in which the metal is dissolved. The anodic reaction involves the dissolution of the metal into the solution, that is, corrosion takes place. Cathodic protection is therefore a method used to control the corrosion of a metal surface by making it a cathode of an electrochemical cell (Corrosionpedia, 2017). Cathodic protection controls the cathodic reaction.

The product of the cathodic reaction depends on the environment. The metal takes up an electrode potential called the "corrosion potential." By moving the potential below the spontaneous corrosion value, using an external source to provide electrons, the rate of dissolution of metal can be slowed down. This is cathodic protection.

The supply of electrons (i.e., the cathodic current) dictates the degree of protection possible. The application of more current follows the law of diminishing returns so far as corrosion is concerned. Apart from being able to protect fully and partially, it is also possible to overprotect the structure. As well as being wasteful, overprotection can be damaging to coatings (Figure 3.4).

Before the construction of the pipeline, the following route survey is required:

- Measurement of the electrical resistivity of the soil environment around the pipeline
- Determination of conditions suitable for anaerobic bacterial corrosion
- Determination of various chemical constituents in the soil environment (chlorides, sulfate, sulfides, bicarbonates)

These route surveys will help in choice of anode materials for cathodic protection.

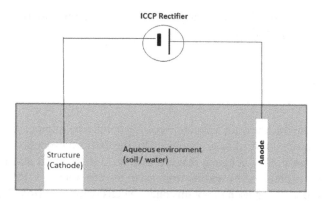

**FIGURE 3.4**   Simple impressed current cathodic protection system.

### 3.3.1   MAIN PARAMETERS OF CATHODIC PROTECTION

### 3.3.1.1   Natural Potential

Natural potential refers to the corrosion potential obtained without the influence of external current when the metal is buried into the soil.

Factors affecting the natural potential are metal material, surface condition and earth quality, and water content. The natural potential for the coated pipeline generally ranges from −0.4 to −0.7 V (CSE), but in the rainy season, the natural potential will be more negative. Cathodic protection (CP) design potential for a new pipeline is −0.55 V on average.

### 3.3.1.2   Minimum Protective Potential

The minimum protective potential is the negative potential with the minimum absolute value required for the complete protection of the metal structure (usually −0.85 V with reference to copper sulfate electrode [CSE]).

### 3.3.1.3   Maximum Protective Potential

The maximum allowable negative potential is slight higher than the potential that can cause the cathodic disbonding (usually −1.25 V with reference to CSE).

This is a process of disbondment of protective coatings from the protected structure (cathode) due to the formation of hydrogen ions over the surface of the protected material (cathode). Disbonding can be intensified by an increase in alkali ions and an increase in cathodic polarization. The degree of disbonding is also reliant on the type of coating, with some coatings affected more than others. Cathodic protection systems should be operated so that the structure does not become excessively polarized, since this also promotes disbonding due to excessively negative potentials. Cathodic disbonding occurs rapidly in pipelines that contain hot fluids because the process is accelerated by heat flow.

### 3.3.1.4   Minimum Protective Current Density

Minimum protective current density refers to the protective current density required to slow the corrosion to the lowest extent or to stop the corrosion process. The common unit is mA/m$^2$.

### 3.3.1.5  Instant Switch-Off Potential

Instant switch-off potential refers to the ground potential obtained 0.2–0.5 s after the sudden break of the external power or sacrificial anode. As there is no external current flow through the protective structure, the obtained potential is the actual polarization potential without IR drop (voltage drop in the medium).

### 3.3.2  SACRIFICIAL ANODE CATHODIC PROTECTION SYSTEM

Potential difference between two different metals of galvanic coupling is the driving force of galvanic corrosion. The strength of corrosion increases as the distance within the galvanic series increases. In a sacrificial anode protection system, there is a certain galvanic distance between the anode and protected structure. The larger the distance is, the more negative potential will be obtained on the protected structure.

Figure 3.5 shows the formation of the sacrificial anode protection system. A complete circuit is formed by connecting the sacrificial anode and the metal to be protected. In this circuit, the sacrificial anode provides effective current to the protected structure, which can be measured.

In general, sacrificial anodes come in three metals: magnesium, aluminum, and zinc. Magnesium has the most negative electro-potential of the three (see galvanic series) and is more suitable for onshore pipelines where the electrolyte (soil or water) resistivity is higher. If the difference in electro-potential is too great, the protected surface (cathode) may become brittle or cause disbonding of the coating.

Zinc and aluminum are commonly used in salt water, where the resistivity is generally lower. Typical uses are for the hulls of ships and boats, offshore pipelines, and production platforms, in salt-water-cooled marine engines, on small boat propellers and rudders, and for the internal surface of storage tanks.

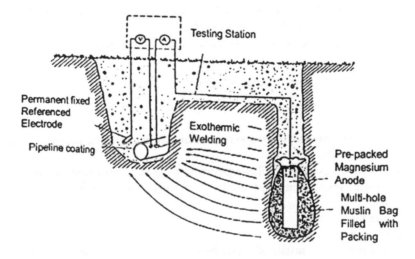

**FIGURE 3.5**  Composition of sacrificial anode protection system. (Source: Tong, Shan. 2015. *Cathodic protection*. Training document, Ghana: Sinopec.)

Requirements for a sacrificial anode system are as follows:

1. The sacrificial anode must have stable and enough negative potential.
2. The anode polarization rate must be small during operation.
3. Theoretical electric capacity of the sacrificial anode material must be enough.
4. Self-corrosion rate of the sacrificial anode must be very small.
5. The sacrificial anode must have uniform activated dissolution during operation.
6. Corrosive products generated during operation of the sacrificial anode must be nontoxic and harmless, with no pollution to the environment.
7. The raw material of the sacrificial anode should be abundant and easy to be processed.

### 3.3.2.1 Advantages
- No external power supply required and are easy to install
- No or less interference to nearby structures
- Less managing workload after commissioning
- Economic for minor engineering
- Uniform distribution of protective current

### 3.3.2.2 Disadvantages
- Not suitable for occasions with high resistivity. Ineffectiveness in high-resistivity environments.
- Consume nonferrous metal.
- Commissioning and debugging operation is complex.
- Well coating is required.
- Protective current is nearly nonadjustable.
- Increased weight on the protected structure, and increased air and water flow on moving structures such as ships.
- Limited current capacity based on the mass of the anode.

### 3.3.3 IMPRESSED CURRENT CATHODIC PROTECTION SYSTEM

The impressed current cathodic protection system is composed of:

1. DC power supply; auxiliary anode; protected structure; environmental media
2. Reference electrode
3. Test station (test post), including current measuring instruments, potential measuring instruments, and resistancemeasuring instruments
4. Cable
5. Insulation device
6. Current loop
   Current loop consists of anode bed, cables, DC power supply (current), protected pipeline, and soil. In this loop, the positive pole of the DC power is connected to the auxiliary anode, and the cathode pole is connected to

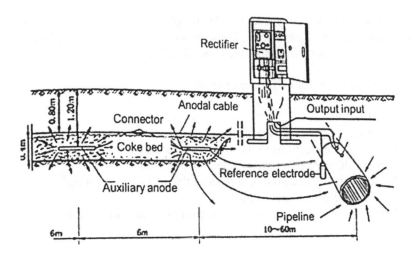

**FIGURE 3.6** Impressed current cathodic protection system. (Source: Tong, Shan. 2015. *Cathodic protection.* Training document, Ghana: Sinopec.)

the protected structure, sending protective current (cathodic protection current) to the protected metal. The effective output current can be tested through a meter in the testing station.

7. Potential loop

Potential loop consists of pipeline, cables, testing station (voltage meter), reference electrode, and soil. Set a permanent reference electrode in environment. Protective potential can be monitored and controlled by the potential meter of the testing station. Potential signal of the protected structure will be fed back to the testing station or a potentiostat so that the output DC can be automatically or manually adjusted to the required protection range (Figure 3.6).

### 3.3.3.1  Advantages

- Output current is continuous and adjustable.
- Large protection range.
- No interference by electrical resistivity of environment.
- Economic for large engineering.
- Long service life of the protection devices.

### 3.3.3.2  Disadvantages

- External power supply required
- Large interference to nearby structures: possibility of contributing to stray current interference on neighboring structures
- Heavy managing workload

### 3.3.4  OFFSHORE CATHODIC PROTECTION

In theory, there are several forms of cathodic protection available, but practically most subsea pipelines are cathodically protected using sacrificial anode bracelets of the half-shell or segmented type attached to the pipeline at regular intervals. For sub-sea structures and platform legs, cathodic protection is by the use of slender stand-off sacrificial anodes or flush-mounted sacrificial anodes which are welded to the structure.

Sacrificial anodes require no source of power.

### 3.3.4.1 Principle

- Difference in galvanic series is used
- To provide the electrons to the protected metals
- To make the surface a state of "electron overload"
- To eliminate the potential difference at the surface of the metal
- To stop the electron movement
- To stop the atoms from being changed into ions that will flow into the earth (Figure 3.7)

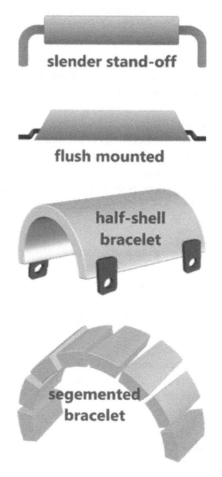

**FIGURE 3.7**   Typical sacrificial anodes used for cathodic protection.

The most common anode configurations are bracelet and stand-off design. However, in some cases, remote anode arrays may be used.

### 3.3.4.2 Anode Design and Attachment

Once the anode current output and weight criteria are both satisfied, the anode shape and method of attachment to the pipe or structure must be designed to be compatible with:

- Pipeline coating materials, dimensions, and application sequence/methods
- Onshore prefabrication, where applicable
- Offshore and subsea construction activities
- Replacement and inspection activities

One of the key points of design is the need to ensure that areas that should be isolated from the cathodically protected steelwork are not allowed to drain current. The most common anode configurations are bracelet and stand-off design. However, in some cases, remote anode arrays may be used.

#### 3.3.4.2.1 Bracelet Anode Shapes

There are two main types of bracelet anode, the twin half-shell and the segmented bracelet. Segment shapes are more easily cast, particularly in large diameters and thicknesses. However, the segmented shape has more complex steel fabrication and hence half-shells are preferred when casting is straightforward. Maximum lengths of half-shells are around one meter, whereas segmented anodes can be cast longer. Anodes can also be "doubled up" or in "clusters" where greater mass is required. Sketches and dimensions of typical anode shapes can be found in manufacturers' literature.

Steel insert material is used for providing an electrical connection point to the pipe as well as holding the anode material together. The mass of the insert material, a cage of welded bars, strips, or plates, must be taken from the gross anode mass to give the net mass ($W_a$). The manufacturer will supply the gross mass of the anode (which includes reinforcements).

Anodes for weight-coated pipelines are usually designed to lie flush with the outer surface of the concrete; otherwise they are exposed to damage during pipe laying. Recessed anodes can be used where the ratio of current output to weight needs increasing. Bracelet anodes for non-weight-coated pipes are normally targeted at each end to ease their passage through handling equipment, stringer, and trenching machine without damage. Bracelet anodes are normally used with small-diameter pipes, that is, 3inch outer diameter and below, and consultation with anode manufacturers should be made before designing a CP system.

When pipelaying requires anode installation to be undertaken "quickly" on the lay vessel before passage through the stringer (e.g., reeling of pipe), hinged bracelets are particularly useful as compared with welded half-shells. The utilization factor can be preserved by coating the inside and edge surfaces of the anode.

Anodes for a piggy-backed line may need to allow for the anodes to be recessed such that the piggyback line can lie, supported, inside the anode. Since large pieces of anode material are effectively removed, the mass must be compensated for by increasing the length and/or thickness.

*3.3.4.2.1.1   Bracelet Anode Attachment*   Attachment of the bracelet must ensure two things:

- A permanent, low-resistance electrical contact between anode and pipe
- Prevention of anode movement, which may break electrical connections during installation

### 3.3.4.2.2   Stand-Off Anodes (Structures)

The design of these anodes is normally more straightforward than for bracelet anodes. For different shapes and sizes, the designer should refer to the manufacturer's data. Anodes for subsea structures are normally placed symmetrically on the structure to give equal load and current distribution. Anodes attached to the top chord should hang vertically down, and those on the top bottom chord should be attached to outside members where damage is more likely to be incurred during installation.

Anodes should be placed using sensible judgment regarding areas of high current demand.

### 3.3.4.2.3   Electrical Continuity

Electrical continuity between the steel and bracelet anodes (on pipelines) or stand-off anodes (for subsea structures) is via the mechanical attachment where the insert material is welded to the pipeline structure or doubler plate. In other cases, for example:

- Piggy-backed lines which are to be protected by the anodes attached to the larger-diameter flowline; or
- Steelwork located near the subsea structure which is protected by the stand-off structure anodes

A short length of electrical cables (normally at least two per anode, for redundancy in case of damage) are run between the anode insert material and the pipe wall. One of the most established methods of attaching cables is Cad welding, but more recently pin brazing has become an established technique for cable attachment.

### 3.3.4.3   Anode Materials

There are several anode materials/alloys in use. Some have specific use and are not suitable for use in all situations of cathodic protection. The three main metals used as sacrificial anodes are magnesium alloy, zinc alloy, and aluminum alloy.

Magnesium alloys are consumed quickly and only have an 8- to 10-year life. Magnesium is often used in soil to protect small electrically isolated structures, such as underground storage tanks and well-coated pipelines.

Aluminum alloys are used for subsea pipelines and have three times the current capacity of zinc; the most widely used being Al-Zn-In alloy (Al-Zn-In sacrificial anodes at different concentrations of zinc and indium). Aluminum can be used for a variety of marine applications.

Zinc cannot be used for buried or unburied pipelines above 50°C but is still used for high-resistivity applications. Zinc is often used in marine and soil environments. They are commonly found on boats.

### 3.3.4.4 Monitoring of Offshore Cathodic Protection System

For monitoring of offshore pipeline or offshore structure cathodic protection systems away from landfalls and offshore platforms, there are two methods that can be used. These methods are described next.

#### 3.3.4.4.1 Remote Permanent Potential Monitoring Devices

The remote permanent monitoring method is normally used offshore for structures in depths below diving range and where ROVs can gain no earthing cable contact. This technique transmits the measured structure to seawater potential via an acoustic transponder (ISO15589-1, 2003).

#### 3.3.4.4.2 Measurement of Structure to Seawater Potential

The reference electrodes used for monitoring marine cathodic protection systems are the silver/silver chloride/seawater electrode and the zinc/seawater electrode (EN12473, 2006).

Measurement of pipe to seawater potential, current density, or anode current output can be achieved through any of the methods stated next.

#### 3.3.4.4.3 Monitoring by Divers

The diving method is expensive and therefore used only for inspection and maintenance work around platforms, wellhead templates, valve assemblage, shallow waters, and so on. Divers may be used where an ROV cannot inspect and for very detailed inspections. Information transmitted by divers is not always reliable and reproducible, therefore such measurement may be carried out with additional aid of a television camera so that the technician on board can record the position and measurement on videotape (Ashworth, 1986).

There are two options for monitoring the potential with the aid of divers:

1. Periodic subsea intervention by divers using hand-held contact probe reference cells to measure structure to seawater potential.
2. The diver holds the reference electrode as close as possible to the structure and gives the position by radio telephone to those on board where the measurement is recorded. This method is only suitable for coated structures.

#### 3.3.4.4.4 Monitoring by Remote Operated Vehicles (ROV)

ROVs are frequently used for pipelines and risers. The risers are usually inspected with platform inspection. There are three methods by which an ROV can be used to measure the pipeline protection potential and anode output. These methods are as follows:

1. Proximity type cell on ROV and hardwire (trailing wire) connected to platform
2. Two reference cells, one attached as above and one "remote" from pipeline.
3. Two reference cells mounted on ROV

Methods (i) and (ii) are hazardous or need frequent calibration, respectively. Method (iii) measures potential as do (i) and (ii) but can calculate current densities applied to pipe and output current densities of anodes. The ROV monitoring method can be used for both uncovered and buried pipelines (Baeckmann et al., 1997). For the purpose of cathodic protection monitoring, proximity type instruments are more accurate than contact probes.

### 3.3.4.4.5   Monitoring by Towed Instruments

Monitoring by towed instruments is similar to ROV technique, where an unpowered "fish" is towed. Although less time-consuming, this is also a less accurate technique. It is suitable for buried pipelines.

Close interval potential surveys provide a nearly continuous plot of the pipeline potential. Towed "fish" or ROVs can be used to carry the monitoring equipment (NRC, 1994). ROVs, which follow the pipeline more closely and consistently, are generally most effective. They also can carry video cameras, which reveal even minor coating defects (on pipelines that are not covered with sediments).

### 3.3.4.5   Criteria for Cathodic Protection

For offshore pipelines or structures, steel is under-protected at potentials below −0.80 V (Ag/AgCl/seawater) and below −0.83 V, under-protection may occur.

### 3.3.5   ONSHORE CATHODIC PROTECTION

Two types of cathodic protection are used for onshore pipelines (CSA Z662 2015):

1. Sacrificial anode (also known as galvanic anode)
2. Impressed current

### 3.3.5.1   Anode Materials

This section details the various anode materials available on the market.

### 3.3.5.1.1   Sacrificial Anode Materials

For onshore cathodic protection applications, two types of sacrificial anodes materials are used with varying composition to suit a particular application in accordance with ASTM B843. These are as shown in Table 3.2.

---

**TABLE 3.2**
**Sacrificial Anode Material Specification**

| Sacrificial Anode Material | Specification |
|---|---|
| Magnesium | High purity (standard) |
|  | Galvomag or Maxmag (high potential) |
| Zinc | High purity |
|  | US Military Specification (US Mil Specification) |

---

Using zinc anodes is restricted to areas where the soil resistivity is less than 5,000 ohm.cm, and magnesium anode uses are normally restricted to areas with soil resistivity less than 10,000 ohm.cm.

High-potential magnesium anodes may be used in soils with higher resistivity due to the greater driving voltage, although the current output will still be restricted.

Sacrificial anodes are normally supplied in a cotton bag filled with a low-resistivity backfill (MESA 2000). This improves the current output of the sacrificial anodes.

Aluminum alloys are normally used for subsea pipelines. For onshore application, aluminum alloys are used for the cathodic protection of pipelines in soils with very low resistivity, less than 1,000 ohm.cm (Figures 3.8 and 3.9).

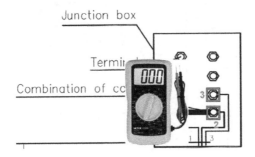

**FIGURE 3.8**   Galvanic CP. (Source: Tong, Shan. 2015. *Cathodic protection*. Training document, Ghana: Sinopec.)

### 3.3.5.1.2   Impressed Current Anode Materials

Impressed current anode materials that may be used for onshore applications are:

1. High silicon cast iron anodes (widely used for impressed current applications [see Figure 3.10])
2. High silicon chromium iron anodes (used where there is possibility of chloride contamination of the ground bed site or anaerobic soil)
3. Graphite
4. Magnetite
5. Platinized titanium (borehole ground beds only)
6. Platinized niobium (borehole ground beds only)

The source of power for an impressed current system could be solar energy battery, wind power generator, transformer-rectifier, thermos batter, and so on.

Consideration should be given to isolation joints or insulating flanges at these interfaces:

1. Onshore to offshore pipelines interface
2. Pipeline to structure/terminal interface (Figure 3.11)

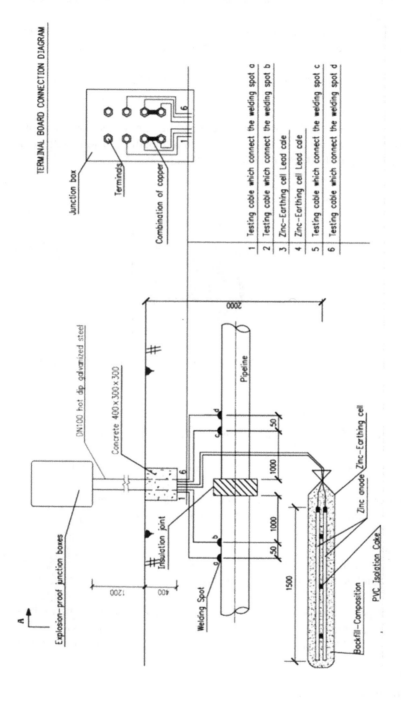

**FIGURE 3.9** Galvanic CP and insulation joint. (Source: Tong, Shan. 2015. *Cathodic protection*. Training document, Ghana: Sinopec.)

**FIGURE 3.10** High-silicon cast iron anode. (Source: Tong, Shan. 2015. *Cathodic protection*. Training document, Ghana: Sinopec.)

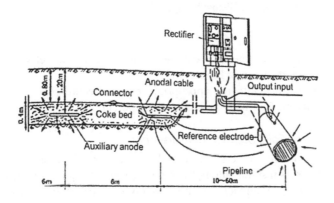

**FIGURE 3.11** Impressed current cathodic protection system. (Source: Tong, Shan. 2015. *Cathodic protection*. Training document, Ghana: Sinopec.)

### 3.3.5.2 Monitoring of Onshore Cathodic Protection System

Cathodic protection monitoring comprises regular inspections performed by pipeline operators to perform measurements at rectifiers and cathodic protection test points along a pipeline (ISO15589-1 2003). The purpose of monitoring is to ensure that the cathodic protection installation remains effective throughout the design life of the structure.

Effective monitoring is by the measurement of structure-to-electrolyte potentials, using a high-impedance voltmeter and a reference cell.

This section is mainly about the various methods used for monitoring an onshore cathodic protection system.

#### 3.3.5.2.1 Natural Potential Measurement

Make sure that cathodic protection is not applied to the pipeline before measurement. For pipeline that has been protected, it is suitable to measure after the power supply is shut down for 24 hours. The copper sulfate electrode is placed on the surface moist soil above the pipe. Make sure the bottom of the copper electrode is well

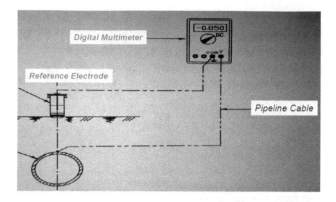

**FIGURE 3.12**   Natural potential (corrosion potential). (Source: Tong, Shan. 2015. *Cathodic protection*. Training document, Ghana: Sinopec.)

contacted with the soil. Connect the voltmeter, pipeline, and copper sulfate electrode following the method shown in Figure 3.12.

Regulate the voltmeter to the appropriate range. Record pipe-to-soil potential value and polarity marked with the name of the potential.

### 3.3.5.2.2   Pipe-to-Soil Potential Measurement

Monitoring of cathodic protection potentials on onshore pipelines is carried out at the test post facilities installed along the pipeline route using a portable copper/copper sulfate reference cell and a high-impedance voltmeter (Figure 3.13).

**FIGURE 3.13**   Pipe-to-soil potential measurement on Ghana National Gas Company pipeline.

*3.3.5.2.2.1   Earth Surface Reference Measurement*   The earth surface reference measurement is used for the open-circuit or closed circuit potential measurement of the pipeline natural potential, protection potential, and sacrificial anode.

- The wiring connection is shown in Figure 3.14.
- Use high-impedance voltmeter. It is recommended to use digital multimeter, such as TD-830.
- The reference electrode is placed on the moist soil surface above the pipe, ensuring it is well connected with the soil. Use the same method every time to reduce error.
- The voltmeter should be adjusted to the appropriate range.

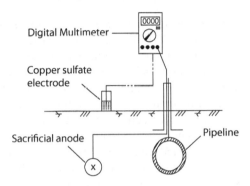

**FIGURE 3.14**   Surface reference. (Source: Tong, Shan. 2015. *Cathodic protection.* Training document, Ghana: Sinopec.)

*3.3.5.2.2.2   Near Reference Measurement*   Near reference measurement is recommended if IR drop is too large. As shown in Figure 3.15, reference electrode is placed close to the pipe wall at about 3–5 cm.

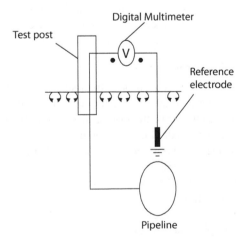

**FIGURE 3.15**   Near reference. (Source: Tong, Shan. 2015. *Cathodic protection.* Training document, Ghana: Sinopec.)

*3.3.5.2.2.3 Outage Measurement*  Outage measurement is used for potential measurement to eliminate the influence of IR drop.

Outage is realized via current interrupter that should be connected in series with current output terminal of cathodic protection.

During non-measuring period, cathodic protection station is energized continuously. During measuring of potential of cathode protection or resistance of coating, cathodic protection station is energized intermittently, supplying every 12 s with an interval of 3s. For the same cathodic protection system, interval power should be supplied simultaneously with the synchrony error no more than 0.1 s. The measured potential value obtained via surface measurement with power on is pipeline protection potential when reference electrode is placed on pipeline.

*3.3.5.2.2.4  Close Interval Potential Surveys (CIPS) or Close Interval Surveys (CIS)*
Close Interval Potential Surveys (CIPS) or close interval surveys (CIS) are used to determine the level of cathodic protection being experienced along the pipeline by measuring the pipe-to-soil potential and hence corrosion protection levels. There are three types of close interval surveys; namely, on/off, depolarized, and on.

A measurement is taken by connecting the high-resistance voltmeter negative lead to the reference electrode (half-cell), and connecting the positive lead to the metal being tested. The reference electrode must contact the electrolyte that is in contact with the metal being tested. In soil and freshwater, a copper/copper sulfate reference electrode should be used; in saltwater, a silver/silver chloride reference electrode must be used. To prevent erroneous readings, the voltmeter used must have a minimum of 10 million ohms input resistance under normal conditions; under rocky or very dry conditions, it should have up to 200 million ohms input resistance.

### 3.3.5.2.3   Remote Monitoring

Remote monitoring units (RMU) are available to remotely monitor the output of transformer-rectifier units and drain-point potentials. Remote monitoring reduces the cost related to conveying engineers to remote sites. Remote monitoring allows wireless transmission of corrosion and cathodic protection data over long distances.

### 3.3.5.3   Criteria for Cathodic Protection

For onshore pipelines/structures, based on a field survey and laboratory test of the soil samples, a decision can be made on the required minimum pipe-to-soil protection criteria. For aerobic soils, the protection criteria would be $-0.85$ V with respect to a $Cu/CuSO_4$ reference electrode. For anaerobic soils containing sulfate-reducing bacteria, the protection criteria would be $-0.95$ V with respect to a $Cu/CuSO_4$ reference electrode (ISO15589-1: 2003).

A pipe-to-soil potential of $-0.85$ volts versus a $Cu$-$CuSO_4$ electrode indicates satisfactory cathodic protection.

The $-0.85$ criteria as well as a potential shift of $-300$ mV versus a $Cu$-$CuSO_4$ electrode is used for bare pipelines. It is also suggested that a $-100$ mV shift of potential in bare pipelines indicates a good degree of cathodic protection.

For high-strength steels with tensile strength in the range of 700–800 MPa, typical protection criteria would be $-1.0$ V with respect to a $Cu/CuSO_4$ reference electrode.

### 3.3.5.4   Protective Potential Value (NACE RP 0169-96, SY/T 0036)
- Energized Condition: −850 mV
- Polarized Potential: *−850 mV*
- Polarized Potential Difference: *−100 mV*

### 3.3.5.5   Test Conditions
- IR drop must be eliminated under energized conditions.
- IR drop can be eliminated by applying power-breaking methods.
- Attenuation process during polarization.
- Probe detection method must be applied when there is stray current interference (Table 3.3).

## TABLE 3.3
## Explanation to the Readings from Cathodic Protection Test

| Voltmeter Reading | Explanation (What the Voltmeter Reading Indicates) |
|---|---|
| Greater than −0.88 volt | These sections of the pipeline are adequately protected. |
| −0.85 volt to −0.88 volt | The pipe structure at those sections still meets the standard for corrosion protection, but there is not much of a safety cushion. We have to monitor those sections of the pipeline with such readings closely to determine the rate at which the voltage is dropping and plan on adding anodes or performing other work on the system in the not too distant future. |
| Less than −0.85 volt | The pipe structure at these sections does not meet the −0.85 volt standard for corrosion protection and is out of compliance with regulatory requirements (refer to Clause 5.3.2.1 of ISO 15589-1). Possible reasons for failure to achieve −0.85 volt is:<br>• Excessively dry soil around the test point.<br>• If the pipe's backfill was damp, but −0.85-volt reading still could not be measured, then there is a need to research the installation procedures to see if we can discover any clues.<br>• Break in a continuity bond or increased resistance between point of connection and point of test due to a poor cable connection.<br>• Deterioration of or damage to the pipeline protective coating.<br>• Reversed connections at the transformer-rectifier, which is a very serious fault that could result in severe damage to the pipeline in a relative short period. |
| −0.4 volt to −0.6 volt | It means the steel pipe has no cathodic protection or that the anodes are completely shot.<br>• There is need for investigation by a corrosion engineer. |

### 3.3.5.5.1   *Steel Casing Pipe Test Station*

When a pipeline is encased in concrete or buried in dry or aerated high-resistivity soil, values less negative than the criteria above may be sufficient.

At casing, the test station usually has four wires, two to the casing and two to the carrier pipeline (Figure 3.16).

Under normal conditions, the carrier pipeline should be at a potential more negative than *−0.85 volts (Negative 850 millivolts) and the casing should be between approximately −0.35 and −0.65 volts* (a difference of between negative 200 to 500 mV).

**FIGURE 3.16**    Typical casing installation.

### 3.3.6 REFERENCE ELECTRODE

A reference electrode is defined as the reversible electrode to measure potentials of other electrodes. There is no net change on its surface over time, and it is always in a balanced state. It will be considered as the electrode that has constant open-circuit potential in similar measuring conditions. The reference electrode is the basic signal source of operation of cathodic protection power supply. Considering maintenance operation and cost, a copper sulfate reference electrode ($Cu/CuSO_4$) is usually used in cathodic protection.

Other standard reference electrodes may be substituted for the saturated copper/copper sulfate reference electrode. Two commonly used reference electrodes are listed here along with their voltage equivalent (at 25°C, [77°F]) to −850 mV, referred to a saturated copper/copper sulfate reference electrode:

- Saturated KCl calomel reference electrode: −780 mV
- Saturated silver/silver chloride reference electrode used in 25 ohm-cm seawater: −800 mV

Reference electrodes are used to measure the oxidizing or reducing power of the working electrode/anode in a cathodic protection system. Cathodic protection reference electrodes are used in locations where a constant, stable reference is required, with constant potential rectifiers and at inaccessible locations. Examples of reference electrodes are Cu/CuSO$_4$ reference electrode and Ag/AgCl reference electrode. The requirements for reference electrodes include small polarization, stability, and long service life.

### 3.3.6.1   Application and Maintenance

- Keep it clean. Cover the plug with a rubber cap when being stored.
- Remain uncontaminated.
- Get certain number of backups in case of loss.
- Get a brand-new one for calibration purposes.

To prevent erroneous potential measurements, the accuracy of the reference electrode (half-cell) must be reliable. A valid (tested) reference electrode must be used to take all potential measurements. Proper maintenance of the half-cell is essential. If the electrolyte in the half-cell is contaminated, or the metallic electrode is contaminated or oxidized, the potential of the cell changes (Guyer, 2014).

- Temperature also affects the potential of the reference cell. There is an increase of approximately 0.9 mV per degree Celsius (0.5 mV per degree Fahrenheit), so a measurement of −0.85 at 21°C (70°F) would read −0.835 at 4°C (40°F), and 0.865 at 38°C (100°F).
- To determine the accuracy of a reference electrode, multiple reference electrodes must be used. In practice, there should be one reference electrode maintained properly, which is not used in the field, to check other reference electrodes against before they are used in the field. This "reference" reference electrode must be properly initiated and stored.

The copper/copper sulfate reference electrode must be properly cleaned and initiated to ensure accuracy. Improper cleaning or initiation can cause significant changes in the potential of the reference (and subsequent errors in all measurements taken). The metal electrode must be cleaned properly, and the electrolyte solution must be prepared properly to ensure accuracy.

Clean the metallic electrode thoroughly using nonmetallic materials: Do not use metallic sandpaper, grinders, emery cloth, wire brushes, knife blades, or any other metallic cleaning method.

The proper way to clean the copper rod is with nonmetallic sandpaper, such as flint paper, and a cloth. All of the copper oxide (green color) must be removed from the electrode, and it should be clean and shiny (no pitting). If the electrode is pitted, the accuracy is questionable and it should be replaced.

The electrolyte must be a fully saturated solution of copper sulfate. The half-cell body must be thoroughly cleaned, then rinsed out several times with distilled water before mixing the solution in the half-cell. There should be approximately one-third the volume of copper sulfate crystals installed in the half-cell, and then the remaining volume should be filled with distilled water. There should not be any copper sulfate crystals in the threaded area of the half-cell. This can be accomplished by slowly adding the distilled water to the threaded area while rotating the half-cell. The proper solution is a deep blue in color, and after vigorous shaking there must still be some copper sulfate crystals that will not go into solution (fully saturated). If the half-cell has been previously used, additional steps are required. All parts should be inspected for cracks or other defects (Guyer, 2014).

To determine the accuracy of a half-cell, use multiple reference electrodes. Using a "reference" reference electrode, measure the difference in potential to the half-cell under test. Use a meter on the millivolt scale, place the two cells cone-to-cone, and measure the potential difference. The potential difference should not be in excess of 5 mV (Figures 3.17 through 3.19).

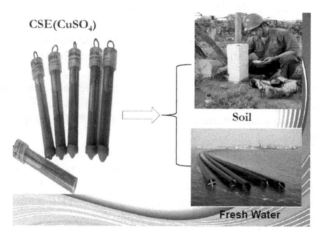

**FIGURE 3.17** Reference electrode. (Source: Tong, Shan. 2015. *Cathodic protection.* Training document, Ghana: Sinopec.)

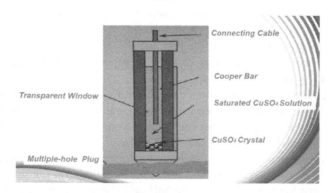

**FIGURE 3.18** CSE(CuSO$_4$) reference electrode. (Source: Tong, Shan. 2015. *Cathodic protection.* Training document, Ghana: Sinopec.)

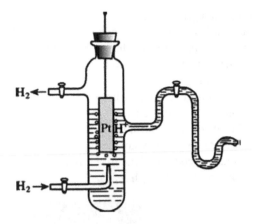

**FIGURE 3.19** Standard hydrogen electrode. (Source: Tong, Shan. 2015. *Cathodic protection.* Training document, Ghana: Sinopec.)

### 3.3.7 GROUNDBED SITE SELECTION AND DESIGN

Ideal impressed current systems use groundbed material that can discharge large amounts of current and yet still have a long life expectancy.

The majority of onshore pipeline cathodic protection systems are the impressed current type and consist of an inert groundbed comprising silicon iron anodes in a backfill of coke breeze, powered by a suitably rated transformer-rectifier unit. While the ideal site for a groundbed is an area of low-resistivity soil somewhere along the pipeline route, the choice of location is often limited by lack of AC supply, environmental constraints, or engineering considerations.

Consequently, the "ideal" site may not be feasible, and a compromise will need to be made. Based on the results of a detailed site survey, horizontal, vertical, or deep well borehole groundbeds may be used to achieve the desired resistance to earth and hence the necessary current output from the transformer-rectifier (Tong, 2015).

Three most common used groundbed types are:

1. Deep well anode groundbed
   - Placement of anodes depends on area of lowest soil resistivity readings
   - PVC casing normally used in the first 25–100 feet for unconsolidated soils, prevent cave- ins
   - Vent line used to displace the corrosion product gases to prevent causing high-resistivity barriers around the anodes; if not placed, could shut down the system over time
   - Coke breeze in deep well systems:
     – The coke breeze is pumped into the well.
     – It must first be mixed with water to form a slurry solution.
     – Then it is pumped into the well starting from the bottom to the top.
     – It must cover all anodes in the well uniformly.

2. Conventional (remote) groundbed

The term "remote," in association with this type of groundbed design, means that the pipeline is outside of the anodic gradient of the groundbed caused by the discharge of the current from the anodes to the surrounding soil.

3. Distributed groundbed
   - Used to protect a limited area of the pipeline.
   - Anodes are generally installed close to the structure.
   - Reduce influence on neighboring structures.
   - Great for bare or ineffectively coated pipelines.
   - Great for areas of congested facilities that could result in electrical shielding (Figures 3.20 through 3.22).

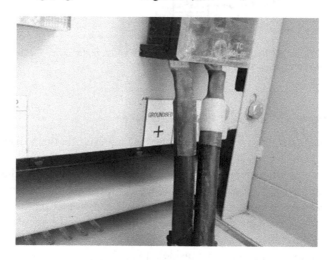

**FIGURE 3.20**  KHV-20A-40V Potentiostat. (Source: Tong, Shan. 2015. *Cathodic protection*. Training document, Ghana: Sinopec.)

### 3.3.7.1  Ground Resistance Measurement of Anode Bed

1. Measurement of grounding resistance of anode ground bed usually adopts three-line measurement.
2. Measuring wirings and instruments are connected following Figure 3.23. Connect blue wire to terminal H, the red wire to terminal S, and the black wire to terminal E.
3. Turn on the power supply. Adjust the function switch to position EARTH3WIRES to short out the measuring wiring.
4. Press SAVE button, displaying the letter K, and then press START button. Displaying the wire resistance value first and then displaying "0.00" means that the instrument has been calibrated, and measurement can start. The instrument should be re-calibrated if the measuring wiring is changed.
5. Insert voltage probe to ground 20 m away from anode bed and current probe 40 m away from anode bed. The wirings of the two probes must be perpendicular to the anode bed. The soil of the inserted position should be moist; otherwise it should be wet by water. The insert depth is around 0.2 m.

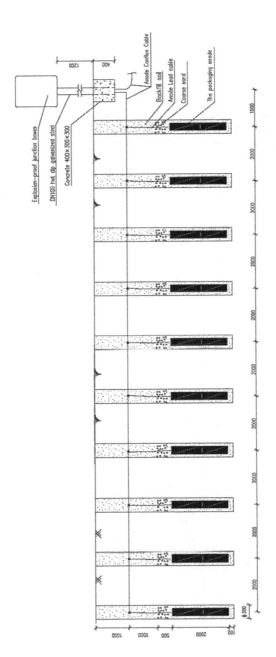

**FIGURE 3.21** Typical installation of auxiliary ground bed. (Source: Tong, Shan. 2015. *Cathodic protection.* Training document, Ghana: Sinopec.)

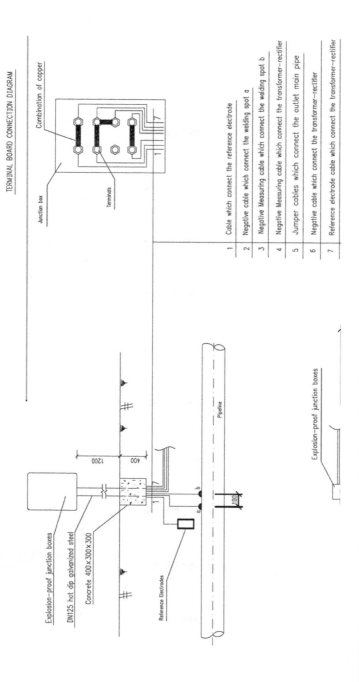

TERMINAL BOARD CONNECTION DIAGRAM

Combination of copper

Junction box

Terminals

| | |
|---|---|
| 1 | Cable which connect the reference electrode |
| 2 | Negative cable which connect the welding spot a |
| 3 | Negative Measuring cable which connect the welding spot b |
| 4 | Negative Measuring cable which connect the transformer–rectifier |
| 5 | Jumper cables which connect the outlet main pipe |
| 6 | Negative cable which connect the transformer–rectifier |
| 7 | Reference electrode cable which connect the transformer–rectifier |

Explosion–proof junction boxes

DN125 hot dip galvanized steel

Concrete 400×300×300

Reference Electrodes

1200

400

Pipeline

Explosion–proof junction boxes

**FIGURE 3.22**  Drain points and measuring cable. (Source: Tong, Shan. 2015. *Cathodic protection*. Training document, Ghana: Sinopec.)

6. Connect the blue wire to the current probe, the red wire to the voltage probe, and the black wire to the anode bed. Meanwhile, disconnect the anode bed from the cathodic protection system.
7. Press the START button. The value displayed is the grounding resistance of the anode bed. Press the SAVE button to store the measured values.
8. Press RCL to read the stored value, if necessary.

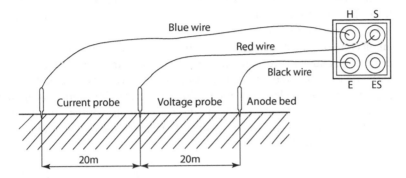

**FIGURE 3.23**   Anode bed ground resistance measurement. (Source: Tong, Shan. 2015. *Cathodic protection*. Training document, Ghana: Sinopec.)

### 3.3.7.2   Ground Resistance Measurement of Sacrificial Anode

This method is suitable for measuring the ground resistance of the grounding body with a diagonal length of less than 8 m.

Disconnect the grounding body and pipeline before measuring. Arrange the electrodes following Figure 3.24 along the direction vertical to the pipeline, with d13 about 40 m and d12 about 20 m or so. Turn the handle of the grounding resistance measuring instrument, making the hand generator achieve the rated speed. Adjust the balance knob until the meter pointer points at the black line. Thus, the value indicated by the black line multiplied by the ratio is the grounding resistance value.

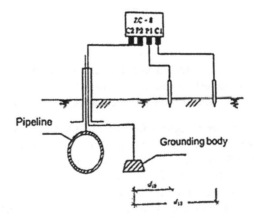

**FIGURE 3.24**   Ground resistance measurement of sacrificial anode. (Source: Tong, Shan. 2015. *Cathodic protection*. Training document, Ghana: Sinopec.)

### 3.3.8 Transformer-Rectifier

A power transformer is an electrical device used to change the magnitude of an AC voltage. A static electrical device transmits electrical energy by converting AC voltage and current between two or more windings at the same frequency according to the principle of electromagnetic induction (Tong, 2015).

A transformer-rectifier is static electrical equipment that is used for altering the AC voltage. By the principle of electromagnetic induction, power is transferred through two or more windings in the same frequency.

The main components of the transformer are the iron core (closed iron core, no air gap, and the main magnetic flux can play the most important role; the leakage flux is very small) and the two windings on the core. The two windings are only magnetically coupled and have no electrical connection. An alternating voltage is applied to the primary winding to generate an alternating magnetic flux of the first and second windings, and an electromotive force is induced in each of the two windings—a narrative capability (Figure 3.25).

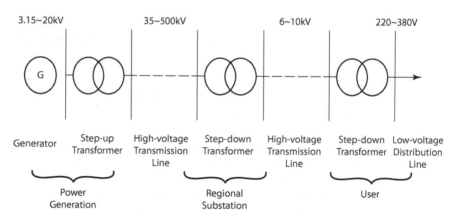

**FIGURE 3.25** Schematic diagram of the transformer. (Source: Tong, Shan. 2015. *Cathodic protection*. Training document, Ghana: Sinopec.)

#### 3.3.8.1 Classification

Transformer-rectifiers for impressed current cathodic protection systems may be three-phase, single-phase, oil-cooled, or air-cooled depending on the location and the intended application. Transformer-rectifiers are classified as follows:

1. Step-up and step-down transformers can be classified based on the function. Step-down transformers are usually applied in factory substations.
2. Based on the number of phases, transformers fall into single-phase and three-phase transformer, the latter of which is mostly used in factory substations.
3. Based on the voltage regulating methods, transformers fall into on-load regulating transformers and off-load regulating transformers (non-excitation regulating transformers), the latter of which are mostly used in factory substations.

4. Based on the winding structures, transformers fall into single-winding auto transformers, dual-winding transformers, and three-winding transformers. The dual-winding transformer is mostly used in factory substations.
5. Based on the winding insulation and cooling methods, transformers fall into oil-immersed transformer, dry-type transformer, and inflatable transformer. And the oil-immersed transformer falls into self-cooling type, air-cooling type, and water-cooling type. The self-cooling type is mostly used in factory substations.

Controlling Transformers: For instrumentation, also includes voltage and current transformers (Figures 3.26 and 3.27).

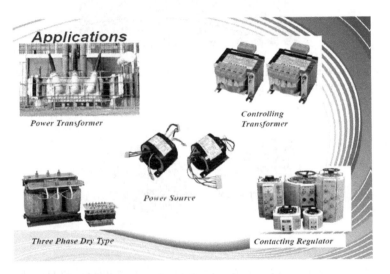

**FIGURE 3.26**   Transformer classification. (Source: Tong, Shan. 2015. *Cathodic protection.* Training document, Ghana: Sinopec.)

**FIGURE 3.27**   Three-phase transformer-oil type. (Source: Tong, Shan. 2015. *Cathodic protection.* Training document, Ghana: Sinopec.)

Three-Phase Transformer, Oil Type: The current power system, transmission, and distribution are all three-phase systems; the three transformers are the most widely used. There are three main points: magnetic circuit structure, connection group, and waveform (Figures 3.28 and 3.29).

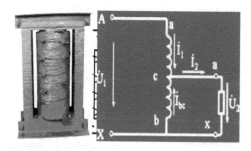

**FIGURE 3.28**   Number of phases. (Source: Tong, Shan. 2015. *Cathodic protection*. Training document, Ghana: Sinopec.)

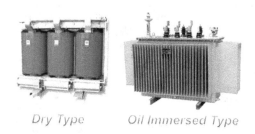

Dry Type                Oil Immersed Type

**FIGURE 3.29**   Cooling type. (Source: Tong, Shan. 2015. *Cathodic protection*. Training document, Ghana: Sinopec.)

Cooling Type Transformer: The same is true for power transformers, which will have a core for the magnetic circuit, which is used as the winding of the circuit. The biggest difference is in "oil" and "dry." That is to say, the cooling medium of the two is different. The former is based on transformer oil (and of course other oils such as β oil) as the cooling and insulating medium, and the latter is air or other gas such as SF6 as cooling medium. Oil change places the body, consisting of iron core and winding, in a fuel tank filled with transformer oil. Drying often casts iron core and winding epoxy resin with epoxy resin (the interlayer insulation is epoxy immersion grease) enveloped. There is also a kind of non-encapsulated type which is used nowadays. The winding is impregnated with special insulating varnish with special insulating paper to prevent the winding or the iron core from being damp (Figure 3.30 and Table 3.4).

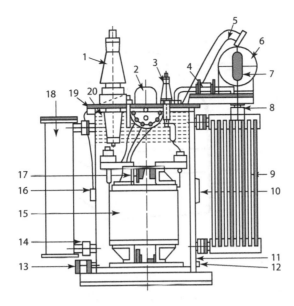

**FIGURE 3.30**  Structure of oil-immersed transformer. 1. High-Voltage Bushing; 2. Tap Switch; 3. Low-Voltage Bushing; 4. Gas Relay; 5. Explosion-Proof Tube; 6. Oil Conservator; 7. Oil Level Gauge; 8. Dehydrating Breather; 9. Heat Radiator; 10. Name Plate; 11. Grounding Bolts; 12. Oil Sampling Valve; 13. Oil Discharging Valve; 14. Valve; 15. Winding; 16. Thermometer; 17. Iron Core; 18. Oil Purifier; 19. Oil Tank; 20. Transformer Oil. (Source: Tong, Shan. 2015. *Cathodic protection.* Training document, Ghana: Sinopec.)

**TABLE 3.4**
**Structure of Oil-Immersed Self-Cooling Type Transformer**

| | | |
|---|---|---|
| Oil-immersed transformer | Case body | Iron core, winding, insulation structure, lead, tap switch |
| | Oil tank | Oil-tank (tank cover, tank wall, tank bottom), accessories (oil discharging valve, oil sampling valve, grounding bolts, name plates) |
| | Cooling device | Heat radiator and cooler |
| | Protective device | Oil conservator, oil level gauge, explosion-proof tube (safety airway), dehydrating breather, thermometer, oil purifier, gas relay |
| | Outlet device | High-voltage bushing, low-voltage bushing |

### 3.3.8.2  Core

The core is one of the most important components of the power transformer, and it is made as a lamination formed by high-permeability silicon steel and steel clamp. The core features two functions (Figure 3.31).

In principle, the iron core is the magnetic circuits that constitute the transformer, which converts the electrical energy into the magnetic energy in the primary circuit where such magnetic energy is converted back to electrical energy in the secondary circuit. Therefore, the iron core is the medium for transferring the energy. In structure it forms the framework of the transformer. Insulation bushings are sheathed on the core limb to firmly support and laminate them.

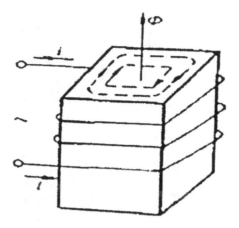

**FIGURE 3.31**  Iron core. (Source: Tong, Shan. 2015. *Cathodic protection*. Training document, Ghana: Sinopec.)

To reduce the hysteresis and eddy loss, the core shall be fabricated into different sizes with the silicon steel plate with the thickness of 0.3–0.5 mm, and insulation painting of 0.01–0.13 mm shall be smeared on the surface of it. After drying, they shall be assembled in an organized manner.

As the resistance of the silicon steel is larger than normal steel plate, eddy loss can be further reduced if the core is fabricated by silicon steel.

### 3.3.8.3  Winding

Winding is the fundamental part of the transformer, which is collectively named as the main body of the transformer together with the iron core. It is the circuit used for establishing a magnetic field and transmitting the power. The types of transformer windings include:

- High-voltage winding
- Low-voltage winding
- Grounding insulation layer (primary insulation)
- Insulating parts between high-low voltage winding

Winding varies in the transformers of different capacity and voltage grade. Concentric types and overlapping types are generally used.

The concentric type is formed by a high-voltage and low-voltage winding on the same core with low-voltage winding placed inside of the high-voltage winding for insulation. For large output capacity transformers, the process of outgoing line of the low-voltage winding is complex, and the winding hereof is usually placed outside of the high-voltage winding. The structure of the concentric type is simple and is easy to be winded; therefore, it enjoys the most extensive application (Figures 3.32 and 3.33).

The role of the winding: The transformer winding constitutes the internal circuit of the device; it is directly connected to the external grid and is the most important component of the transformer, often called the "heart" of the transformer. The change in the number of winding turns can change the voltage. When the winding is assembled with the core, it is both wound into the transformer itself and constitutes an electromagnetic induction system to obtain the required voltage and current. Silicon steel sheets are widely used in medium- and low-frequency transformers and motor cores, especially power frequency transformers. Because silicon steel itself is a kind of material with strong magnetic permeability, it can generate large magnetic induction intensity in the energized coil, which can make the transformer volume smaller and improve the working efficiency of the transformer.

*Features*:

1. Three iron cores are not independent.
2. Three-phase magnetic circuit is related to each other.
3. When three-phase symmetrical voltage is applied, the magnetic fluxes are equal, and the currents are not equal, because the magnetic circuit of the intermediate phase is short and the excitation current is unbalanced, but the actual operation has minimal impact.

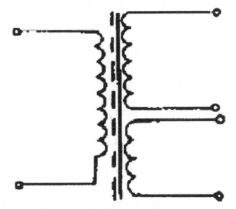

**FIGURE 3.32** Number of windings. (Source: Tong, Shan. 2015. *Cathodic protection.* Training document, Ghana: Sinopec.)

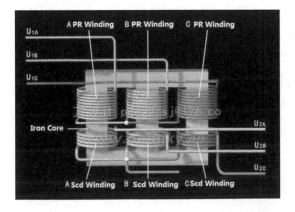

**FIGURE 3.33** Windings. (Source: Tong, Shan. 2015. *Cathodic protection.* Training document, Ghana: Sinopec.)

### 3.3.9 ANODE BACKFILLING

Sacrificial anodes are normally supplied in a cotton bag filled with a low-resistivity backfill (MESA 2000). This improves the current output of the sacrificial anodes. Galvanic anodes must not be backfilled with coke as with impressed current anodes.

The anode used for cathodic protection is not in direct contact with the soil in which it is buried. The reason is that the soil contains many minerals and other chemicals that might affect the anode and therefore decrease its effectiveness. One of the harmful effects that might be caused by minerals is the buildup of high-resistance films on the surface of the anode, thus hampering its conductivity.

In addition, we want the anode to be uniformly consumed and give its maximum efficiency. That is why special backfills are used depending on the particular environment, application, and the anode's material. The prime purpose of using the backfill is to reduce electrical resistivity. This provides a lower anode-to-earth resistance and greater current outputs in cases where the surrounding soil is of high resistivity (Corrosionpedia, 2017).

Therefore, the backfill restricts the formation of surface films and prevents electroosmotic dehydration and acts to provide uniform current delivery and uniform material consumption. The latter is mainly due to the presence of gypsum in the backfill, while bentonite and kieselguhr maintain moisture.

The backfill serves three basic functions:

- It decreases the anode-to-earth resistance by increasing the anode's effective size.
- It extends the system's operational life by providing additional anode material.
- It provides a uniform environment around the anode, minimizing deleterious localized attack.

Some common backfill materials include:

- Coal coke breeze
- Petroleum coke breeze
- Bentonite clay
- Gypsum
- Sodium sulfate
- Kieselguhr (Figure 3.34 and Table 3.5)

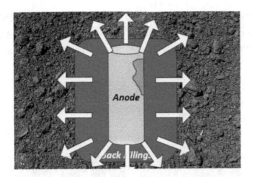

**FIGURE 3.34** Backfilling. (Source: Tong, Shan. 2015. *Cathodic protection*. Training document, Ghana: Sinopec.)

**TABLE 3.5**
**Backfilling for Anode Material**

| | Mass Percentage | | | |
| Anode Type | Gypsum Powder | Bentonite | Industrial Na$_2$SO$_4$ | Applicable Soil Resistivity Ω.m |
|---|---|---|---|---|
| Magnesium | 50 | 50 | – | ≤20 |
| | 75 | 20 | 5 | >20 |
| Zinc alloy | 50 | 45 | 5 | ≤20 |
| | 75 | 20 | 5 | >20 |

*Source*: Tong, Shan. 2015. *Cathodic protection*. Training document, Ghana: Sinopec.
*Note*: molecular formula of gypsum powder is CaSO$_4$.2H$_2$O.

### 3.3.9.1 Backfilling Selection

**TABLE 3.6**
**Backfilling Selection**

| Anode Type | Soil Resistivity (Ω.m) |
|---|---|
| Magnesium alloy | 15~150 |
| Zinc alloy | < 15 |

### 3.3.10 AUXILIARY FACILITIES OF CATHODIC PROTECTION

### 3.3.10.1 Insulation Device

An insulating device is used to protect good electrical insulation between a protected structure and a non-protected structure. It includes an insulating joint, an insulating flange, an insulation sleeve, and so on. Its functions are as follows:

1. To limit the protective current within effective protection range

2. To help to prevent or cut off the interference of stray current
3. To satisfy the design requirement of cathode protection station
4. To make it convenient to inspect and manage the protected system (Figure 3.35)

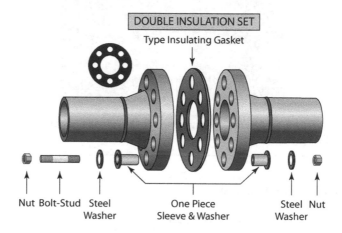

**FIGURE 3.35** Internal structure of insulation joint. (Source: Tong, Shan. 2015. *Cathodic protection*. Training document, Ghana: Sinopec.)

### 3.3.10.1.1  Insulation Performance Test of Insulation Flange

Leaking of the insulation flange can be measured via measuring the pipe-to-soil potential of both sides of the insulating flange. If the potential of the protected side is protection potential with that of non-protected side natural potential, the insulation flange is normal.

1. Potential measurement
   - Potential measurement is used to judge the insulation performance of the insulating flange installed on the pipeline.
   - Before cathodic protection is put into use, measure the potential to the ground $V_{a1}$ of a point a (the non-protected side of insulation flange); make the ground potential $V_b$ of the protected side of the insulation flange reach the protection value 0.85–1.5 V, then measure the potential to ground $V_{a2}$ of point a. If $V_{a2}$ and $V_{a1}$ are basically the same, the insulation performance of the insulating flange is good; if $V_{a2} > V_{a1}$ and $V_{a2}$ is close to $V_b$, the insulating performance of insulating flange is bad.
2. Leaking resistance measurement
   - This method is conducted when the insulation performance is measured and suspicious using potential measurement.
   - Leaking resistance measuring procedures are as follows:
     - Connect measuring circuit well. The horizontal distance between a and b must be more than D, and the distance between b and c should be 30 meters.

- Adjust the output current $I_1$ of mandatory power supply E, making protected potential value each reach cathodic protection value.
- Measure the potential difference $\Delta V$ between d and e with digital multimeter.
- Measure the inner circuit $I_2$ of the pipeline between b and c.
- Read cathodic protection current $I_1$ provided by mandatory power supply.
- Leakage resistance of insulating flange is calculated according to the following formula:

$$RH = \frac{\Delta V}{I_1 - I_2} \qquad (3.3)$$

where:
RH = leakage resistance of insulating flange
$\Delta V$ = potential difference on both sides of the insulation flange
$I_1$ = the output current of mandatory power supply
$I_2$ = inner current of the pipeline between b and c

Leakage percentage of insulating flange is calculated according to the following formula:

$$\text{leakage percentage} = \frac{I_1 - I_2}{I_1 \times 100\%} \qquad (3.4)$$

If the test results $I_1 > I_2$, leakage resistance of insulating flange can be considered infinity. The leakage percentage is zero, and the insulating performance of the insulating flange is good (Figure 3.36).

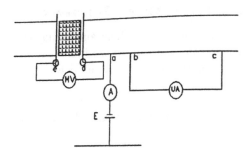

**FIGURE 3.36** Leaking resistance measurement. (Source: Tong, Shan. 2015. *Cathodic protection*. Training document, Ghana: Sinopec.)

### 3.3.10.2 CP Measuring Devices

#### 3.3.10.2.1 Test Posts

A CP system cannot exist without a test post that is used to test the system effectiveness. The following standards shall be conformed when installing a test post:

1. A potential test post shall be installed every 1 km.

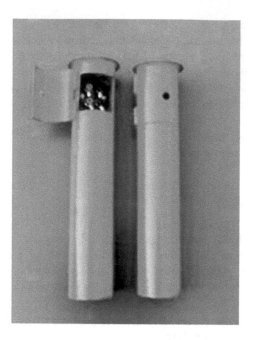

**FIGURE 3.37** Test post. (Source: Tong, Shan. 2015. *Cathodic protection*. Training document, Ghana: Sinopec.)

2. A current test post shall be installed every 5–8 km.
3. A current test post shall be installed at both sides of the large rivers where the pipeline crosses (Figure 3.37).

### 3.3.11 SATISFYING THE CURRENT OUTPUT REQUIREMENT

In order to decrease the amount of sacrificial anode material required for continual protection, submarine pipelines and subsea installations are almost invariably coated with an anticorrosion system. The coating deteriorates with time, exposing an increasing area of bare metal, so that maximum current output by the cathodic protection system occurs at the very end of the installation design life. The final maximum current output requirement of the cathodic protection system is:

$$I_c = A_c \times i_c \times f_c \qquad (3.5)$$

where:
$A_c$ = individual surface areas (m²) of each CP unit
$i_c$ = design current density (A/m²)
$f_c$ = coating breakdown factor (if applicable)

Keys to obtaining enough cathodic protection current are:

1. Determine amount of current required: Theoretical calculations based on coating quality and environment, or performance of current requirement testing.

2. Calculate output expected from anode and determine number of anodes required (Figure 3.38).

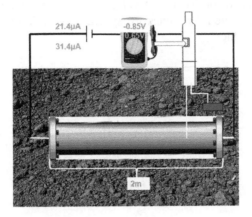

**FIGURE 3.38** Current requirement testing. (Source: Tong, Shan. 2015. *Cathodic protection*. Training document, Ghana: Sinopec.)

### 3.3.12 DESIGN OF OFFSHORE CATHODIC PROTECTION SYSTEM

This section covers the design of offshore/subsea cathodic protection (CP) systems for pipelines and structures. The aim is to provide a method of calculating the required protection and to point out possible problem areas.

The cathodic protection system design is in accordance with DnV-RP-B401, CP 1021, and MTO publications. DnV-RP-B401 is commonly used for North Sea cathodic protection design and is mandatory in the Norwegian sector. CP 1021 is a more general code of practice and is of less use in design calculations.

#### 3.3.12.1 Data Required

The data required for cathodic protection design are:

1. Seawater resistivity: Normally, this information should be provided as project-specific data. However, RP B401 gives some typical values.
2. Seabed soil resistivity: Values are provided in DnV RP B401. Since a range is suggested, if uncertainty exists, the more conservative (higher) value should be used. Field resistivity measurements will be necessary at landfalls where resistivity is much higher.
3. Anode potential: The open-circuit potential of AL-Zn-In alloys is usually taken to be −1.1 V versus Ag/AgCl at ambient temperature. Further information can be obtained from DnV RP B401 and manufacturers' literature. Anode potential decreases (i.e., becomes less negative) linearly with increase in temperature and is expected to be a minimum of −1.0 V at temperatures above 70°C (manufacturers' data).
4. Current capacity ($C_a$): AL-Zn-In anodes have a current capacity of approximately 2650 Ah/kg at ambient temperature in seawater. DnV RP B401

gives the range of typical anodes. Manufacturers' data also gives some values. Current capacity decreases with increased temperature of the pipe. There is some inconsistency of values quoted. The anode manufacturer should be contacted for values for specific alloys. Typical values are as shown in the table.

| Temperature | Mud | Seawater |
|---|---|---|
| 30°C | 2270Ah/kg | 2670Ah/kg |
| 40°C | 2060Ah/kg | 2630Ah/kg |
| 50°C | 1750Ah/kg | 2570Ah/kg |
| 60°C | 1300Ah/kg | 2510Ah/kg |
| 70°C | 1040Ah/kg | 2380Ah/kg |
| 80°C | 890Ah/kg | – |
| 100°C | 800Ah/kg | – |

In the absence of other data, the more conservative values above for mud should be used. If data is given in terms of consumption rate ($E_a$), this is the reciprocal of current capacity, $C_a$ (adjusting the units as necessary).

5. Anode utilization factor (u): This is the fraction of anode material consumed when the remaining material cannot deliver the current required. Values are given in DnV RP B401 for various shapes of anode. For bracelet anodes, it is 0.75 to 0.80.

6. Anode half-shell gap (G): This varies with the size of pipe. Anode manufacturer's data or project drawings give typical values. The gap is usually minimized consistent with the ability to join the half-shells and attach them to the pipe.

7. Anode dimensions ($L_a$, $t_a$): These are based on manufacturers' tolerances. Bracelet anode limits are 1 m long maximum and 40 mm thick minimum. Anodes at the upper limit of length must have a greater thickness. Manufacturers have standard sizes within these limits and the length to thickness ratio is limited by the need to maintain casting quality. For concrete coated pipelines, the anode thickness is normally the same as the concrete thickness ($t_c$) so that the anodes lie flush with the weight coat.

8. Steel potential ($V_p$): Values are given in RP B401, which also explains when to use which limits.

9. Coating breakdown factor: The coating breakdown values are given in DnV RP B401. These figures do not reflect the high-performance coatings such as fusion-bonded epoxy (FBE) and neoprene, where breakdown percentages would be lower.

10. Anode material density ($d_a$): See manufacturer's data. Typical value for an Al-Zn-In anode is 2660 kg/m$^3$.

11. Current density: Current density is given in DnV RP B401 for bare steel in various climates and conditions. The values given include some where coating breakdown has already been accounted for.

12. Current density temperature factor: A current density correction factor must be applied where temperatures exceed 25°C. This factor varies according to conditions as described in various parts of DnV RP B401.
13. Structure parameters: All pipelines to be protected require information on length, nominal outside diameter, corrosion coating thickness and concrete coating thickness values. All dimensions of other steelwork to be protected will be needed to calculate surface areas.
14. Temperature of product/fluids: For flowlines, it is necessary to have a temperature profile for each line which is to be cathodically protected. Temperatures below 25°C can be considered as 25°C.
15. Joint spacing: Pipeline anode spacing may be expressed as a distance or as joint spacings (i.e., pipe lengths) where one joint spacing equals 12.2 m. The maximum joint spacing before current distribution becomes suspect is 12, although 8 is a more suitable spacing where low current demand is required. For hot lines, small spacings will be required, e.g., as low as three (3) joints (36.6 m). For ease of installation and low cost, this spacing is undesirable and should only be used for short lengths of pipe.

### 3.3.12.2   Design Procedure

**1) Surface area calculation**

For each CP unit, surface areas to receive CP should be calculated separately for surfaces with and without a coating system and for surfaces affected by other parameters (e.g., surface temperature) which influence the CP current demand.

**Area of coated steel pipeline**

The area of coated steel for a pipeline is calculated as:

$$A_c = \pi \times d_0 \times l \qquad (3.6)$$

where:

$A_c$ = individual surface areas (m²) of each CP unit
D = outside diameter of pipeline (m)
L = length of pipeline or length of pipeline section (m)

**2) Current demand calculations**

$$I_c - A_c \times i_c \times f_c \qquad (3.7)$$

where:

$A_c$ = individual surface areas (m²) of each CP unit
$i_c$ = design current density (A/m²)
$f_c$ = coating breakdown factor (if applicable)

The coating breakdown factor, fc, describes the anticipated reduction in cathodic current density due to the application of an electrically insulating coating. When fc = 0, the coating is 100 percent electrically insulating, thus decreasing the cathodic current density to zero. fc = 1 means that the coating has no current reducing properties:

$$f_c = a + b \times t \qquad (3.8)$$

where, t (years) is the coating age and a and b are constants that are dependent on coating properties and the environment. Table 10-4 of DnV RP B401 gives the recommended constants a and b for calculating the paint coating breakdown factors.

### 3) Current drain calculation

For buried surfaces of mudmat, skirts, and piles, current drain = 0.020A/m². Current drain of 5A per well casing shall be included in current drain calculation. For any current drain to the steel armoring of flexible pipelines, 0.0005A/m² (based on external pipe surface) is recommended.

### 4) Selection of anode type

Anode types include:

- Stand-off anode
- Flush-mounted anode
- Bracelet anode

These maybe specified based on sea current drag and interference with subsea interventions.

### 5) Anode mass calculation

The total net anode mass (Ma) is calculated as:

$$M_a = \frac{I_{cm} \times t_f \times 8760}{u \times \varepsilon} \qquad (3.9)$$

where:
  $M_a$ = total net anode mass (kg)
  $t_f$ = design life (yrs)
  $I_{cm}$ = current demand (A)
  8760 refers to hours per year
  $u, \varepsilon$ are selected
  $u$ = utilization factor

Table 10-8 of DnV RP B401 gives the recommended anode utilization factors for CP design calculations:

$$\varepsilon = \text{design electrochemical capacity} \left( \text{Ah} / \text{kg} \right).$$

$\varepsilon$ is selected from Table 10-6 of Annex A of DnV-RP-B401.

Table 10-6 of DnV RP B401 gives the recommended design electrochemical capacity and design closed circuit potential for anode materials at seawater ambient temperatures.

## 6) Number of anodes:

$$I_c = N \times I_a = \frac{N\left(E_c^\circ - E_a^\circ\right)}{R_a} = \frac{N \times \Delta E^\circ}{R_a} \tag{3.10}$$

where:

$N$ = number of anodes

$E_a^\circ$ = design closed circuit potential of the anode material (V)

$R_a$ = anode resistance (ohm)

$E_c^\circ$ = design protective potential (V)

$E_c^\circ = -0.80$ for subsea pipelines

$E_c^\circ = -0.85V$ or $-0.95V$ for onshore pipelines

$I_a$ = individual anode current output (A)

$\Delta E^\circ$ = design driving voltage (V)

$I_c$ = current demand (A)

*Alternative method of calculating number of anodes (N):*

$$I_c = A_c \times i_c \times f_c \tag{3.11}$$

The area of coated steel for a pipeline is calculated as:

$$A_c = \pi \times d_0 \times l \tag{3.12}$$

Also from Ohm's Law, for current output of anode at end of life:

$$I_c = \frac{V}{R_a} \times N \tag{3.13}$$

Combining the three equations above (3.11–3.13), a relationship between Ra and N is obtained:

$$N = \frac{R_a \times \left(A_c \times i_c \times f_c\right)}{V} \tag{3.14}$$

Equation (3.14) is applicable to all coated subsea pipelines and structures that are to be protected by any number of anodes.

where:

$$V = V_p - V_a = \text{driving voltage}$$

$V_p$ = positive potential limit for adequate pipeline protection

$V_a$ = anode material open-circuit potential

## 7) Individual anode current capacity:

$$C_a = M_a \times \varepsilon \times u \tag{3.15}$$

where:
    $C_a$ = individual anode current capacity (A.h)
    $M_a$ = net mass per anode (kg)

## 8) The total current capacity for a CP unit with N anodes is calculated as:

$$C_{atot} = N \times C_a \tag{3.16}$$

## 9) Check the following conditions are met:

$$C_{atot} = N \times C_a \geq I_{cm} \times t_f \times 8760$$

$$I_{atoti} = N \times I_{ai} \geq I_{ci}$$

$$I_{atotf} = N \times I_{af} \geq I_{cf}$$

## 10) Calculation of anode resistance
### *Long slender stand-off*
*If L≥4r*

$$R_a = \frac{\rho}{2\pi L}\left(\ln\frac{4L}{r} - 1\right) \tag{3.17}$$

### *Short slender stand-off*
*If L<4r*

$$R_a = \frac{\rho}{2\pi L}\left[\ln\left\{\frac{2L}{r}\left(1+\sqrt{1+\left(\frac{r}{2L}\right)^2}\right)\right\} + \frac{r}{2L} - \sqrt{1+\left(\frac{r}{2L}\right)^2}\right] \tag{3.18}$$

### *Long flush mounted*

if $L \geq 4 \times$ width, or $L \geq 4 \times$ thickness

$$R_a = \frac{\rho}{2 \times S} \tag{3.19}$$

### *Short flush-mounted, bracelet, and other types*

$$R_a = \frac{0.315 \times \rho}{\sqrt{A}} \tag{3.20}$$

The equations (3.17–3.18) are valid for anodes with minimum distance 0.30 m from protection object. For anode to object distance less than 0.30 m but minimum 0.15 m, the same equation may be applied with a correction factor of 1.3. The equations (3.17–3.19) are applicable to non-cylindrical anodes: $r = c / 2\pi$, where c (m) is the anode cross-sectional periphery (refer to Table 10-7 of DnV-RP-B401).

### 3.3.12.3   Optimizing Design Calculations

The results of anode calculations need to be optimized to ensure that the most effective and economical design is achieved. The optimum size of the anode must meet a minimum mass requirement to supply the required current and exceed this mass by as little as possible. The size of the anode must also be one that is available from manufacturers in practice. The areas in which it is necessary to optimize design are:

- Anode length versus anode spacing
- Anode spacing versus temperature along the pipeline

Computers best carry out these calculations, and there is a cathodic protection calculation module in PIPECALC.

It is necessary to be aware of simplifications that may be used in computer calculation methods. For example:

- Assuming a cylindrical-shaped anode where there may be tapered ends or recesses for piggyback lines
- Ignoring the anode material loss due to reinforcing bars

Hand calculations may be necessary to allow for these factors.

### 3.3.13   DESIGN OF ONSHORE CATHODIC PROTECTION SYSTEM

The soil conditions (i.e., current density and soil resistivity) dictates the type of cathodic protection system to use. If the soil resistivity is less than 5000 ohm-centimeters and the current density requirement is less than 1 milliampere per square foot, a galvanic or sacrificial anode system should be used. On the other hand, if the soil resistivity and/or current density requirement exceed the above values, an impressed current system should be used.

In order to calculate the requirements for an onshore CP system, it is necessary to know the surface area of the pipework to be protected and the coating system that is to be used. The choice of coating system will dictate the maximum negative potential that may be applied to the pipeline, which will in turn determine the maximum spacing between the CP installations. The pipe surface area (and coating system) will determine the amount of current required to cathodically protect the pipeline over its design life (and hence the rating of the transformer-rectifier units for impressed current systems).

In the case of a proposed onshore pipeline, a survey of the route is normally carried out to record such things as soil resistivity, particular features that may create potential corrosion hazards (e.g., seawater inlets, electrified rail crossings, major river crossings). Soil samples would also be taken at various locations to determine the presence of sulfides, chlorides, carbonates, and sulfate-reducing bacteria.

### 3.3.13.1   Impressed Current Cathodic Protection System Design

There are 10 steps involved in the design of the impressed current cathodic protection system:

1. Examine the soil resistivity. This will help in the design calculations and location of anode groundbed.
2. Review the current requirement test.
3. Select the anode type, sizes and specification. The commonly used anode material is high silicon cast iron anode (see paragraph 3.3.5.1.2).
4. Theoretical current calculation:

$$I = \text{Area} \times \text{current density} \times \% \, \text{Bare} \qquad (3.21)$$

For pipelines,

$$\text{Area} = \pi \times d \times L \qquad (3.22)$$

Current density $= 2 \text{mA/ft}^2$ (bare steel)

5. Calculate the number of anodes

$$W = \frac{\text{anode consumption rate} \times \text{design life} \times \text{required current}}{\text{Utilization factor}} \qquad (3.23)$$

a.   Number of anodes required based on anode consumption rate

$$N = \left( \frac{Y \times I \times C}{W} \right) \qquad (3.24)$$

b.   Number of anodes based on weight (lb)

$$N = \frac{\text{total weight of anode material}}{\text{weight of one anode}} \qquad (3.25)$$

c.   Number of anodes based on current output

$$N = \frac{\text{total current output} \left( A \right)}{\text{Output of one anode} \left( A \right)} \qquad (3.26)$$

$$\text{Current output of one anode} = \frac{\text{Driving potential of the anode} \left( V \right)}{\text{Resistance of the anode} \left( \text{ohms} \right)} \qquad (3.27)$$

where:
   N = number of anodes
   Y = design life in years
   I = total current required in amperes
   C = anode consumption rate in kg/A-yr
   W = weight of a single anode in kg

6. Calculate the groundbed resistance.

Use formula based on groundbed types/configurations. The groundbed is classified as:
- Single vertical (deep anode)
- Multiple vertical
- Single horizontal
- Multiple horizontal

a. Calculating anode resistance to earth – single vertical anode:

$$R_v = \frac{0.00521\rho}{L} \times \left( \ln \frac{8L}{d} - 1 \right) \tag{3.28}$$

where:

$R_v$ = resistance to earth of single vertical anode-to-earth (ohms)
$\rho$ = soil resistivity (ohm-cm)
$L$ = anode length (ft)
$d$ = anode diameter (ft)

b. Calculating anode resistance for multiple vertical anodes.
Multiple vertical anodes in parallel (using Erling D. Sunde formula):

$$R_v = \frac{0.00521\rho}{NL} \times \left( \ln \frac{8L}{d} - 1 + \frac{2L}{S} \ln 0.656N \right) \tag{3.29}$$

where:

$R_v$ = resistance-to-earth (ohms), of the vertical anodes connected in parallel
$\rho$ = soil resistivity (ohm-cm)
$L$ = anode length (ft)
$d$ = anode diameter (ft)
$S$ = spacing between anodes (ft)
$N$ = number of vertical anodes in parallel

c. Calculating anode resistance single horizontal anode (using H. B. Dwight formula):

$$R_H = \frac{0.00521\rho}{L} \times \left( \ln \frac{4L^2 + 4L\sqrt{S^2 + L^2}}{dS} + \frac{S}{L} - \frac{\sqrt{S^2 + L^2}}{L} - 1 \right) \tag{3.30}$$

where:

$R_H$ = resistance-to-earth (ohms), of the vertical anodes
$\rho$ = soil resistivity (ohm-cm)
$L$ = horizontal anode length (ft)
$d$ = anode diameter (ft)
$S$ = twice anode depth (ft)

7. Determine the rectifier voltage size.
   a.  Determine the total circuit resistance:

$$R_{Total} = R_{groundbed} + R_{structure} + R_{cable} \qquad (3.31)$$

$$R_{structure} = R_{coating} \,/\, \text{Area} \qquad (3.32)$$

$$R_{structure} = R_{coating} \,/\, n \qquad (3.33)$$

where:
$R_{structure}$ = resistance of structure to earth (ohms)
$R_{coating}$ = effective coating resistance
n = surface area (ft$^2$)

Cable resistance, $R_c$, is usually negligible unless traversing really long distances:

$$R_c = R_{cable} \times L_{cable} \qquad (3.34)$$

where:
$R_{cable}$ = standard resistance of cable
$L_{cable}$ = length of cable used

$$L_{cable} = L_{negative} + L_{pos(1)} + 1/2 L_{pos(2)**} \qquad (3.35)$$

** = average current in $L_{pos(2)}$ is ½ total current because of parallel anodes

   b.  Determine the rectifier voltage size
       Use Ohm's Law,

$$V = IR \qquad (3.36)$$

where:
I = current required
R = total circuit resistance

$$R_{Total} = R_{groundbed} + R_{structure} + R_{cable} \qquad (3.37)$$

8. Select the rectifier. Choose the rectifier based on the results of equation (3.36). There are many commercially available rectifiers, so choose a rectifier that satisfies the minimum requirement of (I) and (V) in equation (3.36).
9. Calculate the cost using NACE standard RP-02.
10. Prepare plans and specifications.

### 3.3.13.1.1    Design Examples

Design an impressed current system to protect a buried pipeline coated with fusion-bonded epoxy (FBE) using the following information:

- Horizontal anodes 6 ft below ground with 20 ft spacing
- Anode material: High silicon cast iron (C=0.3 Kg/[A*yr])
- Anode dimensions with backfill: 75 mm dia. × 1500 mm, weight=50 Kg
- Pipeline length: 999.99 m and 508 mm in diameter
- The anode-to-soil resistance is 0.5 ohm
- Neglect cable resistivity
- Soil resistivity ($p$): 75,300 ohm-cm
- Required current density is 0.2 mA/m² for FBE.
- Design life of 10 years

### 3.3.13.1.1.1    Computations

1. Theoretical current calculation

$$I = \text{Area} \times \text{current density} \times \% \text{Bare}$$

For pipelines, $\text{Area} = \pi \times d \times L$
  *Area of steel (A$_s$):*

$$A_s = \pi \times d \times L$$

$$A_s = \pi \times 0.508 \times 999.99 = 1596.12 m^2$$

$$A_s = 1596.12 m^2$$

*Current requirement:*

$$I = I' \times A_s$$

$$I = \left(0.2 \ mA / m^2\right) \times 1596.12 = 319.224 \ mA$$

$$I = 319.224 \ mA = 0.319 \ A$$

where:
  I = current requirement in amperes
  I' = required current density
  As = area of steel to be protected (m²)

2. Calculate the number of anodes

$$W = \frac{\text{anode consumption rate} \times \text{design life} \times \text{required current}}{\text{Utilization factor}}$$

Number of anodes required based on anode consumption rate:

$$N = \left( \frac{Y \times I \times C}{W} \right)$$

$$N = \frac{10 \text{yr} \times (0.319 \text{A}) \times (0.3 \text{Kg} / (\text{A*yr}))}{50 \text{kg}} = 0.019 \text{ Anodes}$$

$$\therefore N = 0.019 \cong 1 \text{ Anode}$$

Therefore, use one (1) anode
   where:
      N = number of anodes
      Y = design life in years
      I = total current required in amperes
      C = anode consumption rate in kg/A-yr
      W = weight of a single anode in kg

  3. Calculate resistance

$$R_H = \rho \times \frac{F}{483}$$

For single horizontal anodes,
   F is called adjusting factor (F=1 for single anode)

$$R_H = 75300 \times \frac{1}{483} = 155.9006 \Omega$$

$$R_H = 155.9006 \Omega$$

Total resistance, R

$$R = R_H + \textit{anode to soil resistance} = 155.9006 + 0.5 = 156.4006 \Omega$$

$$R = 156.4006 \Omega$$

where:
  $R_H$ = resistance-to-earth (ohms), of the vertical anodes
  $\rho$ = soil resistivity (ohm-cm)

  4. Calculate the rectifier voltage size needed

$$V = I \times R$$

where:
  I = current required
  R = total circuit resistance

$$V = 0.319224 \times 156.4006 = 49.927 \ V$$
$$V = 49.927 \ V$$

**Hence, use a DC power with a minimum current supply of 320mA, and a minimum voltage of 50 V.**

### 3.3.13.2 Sacrificial Anode Cathodic Protection System Design

The design of the sacrificial anode system follows the following nine steps:

1. Examine the soil resistivity: to minimize anode-to-electrolyte resistivity; choose site of lowest resistivity for anode location. However, if the soil resistivity does not vary significantly along the project area, then the average soil resistivity will be used for the design.
2. Select anode: As indicated in paragraph 3.3.5.1.1, sacrificial anodes used onshore are usually magnesium or zinc anodes. Examine data from commercially available anodes, and choose the most economical one based on design. Anode specification should include anode weight, anode dimensions, package dimensions, and anode driving potential.
3. Calculate the net driving potential of the anodes.
4. Calculate the number of anodes required to meet groundbed resistance limitations. The total resistance (RT) of the galvanic circuit is given as:

$$R_T = R_a + R_w + R_c \tag{3.38}$$

where:
$R_a$ = anode-to-electrolyte resistance
$R_w$ = anode lead wire resistance
$R_c$ = the structure-to-electrolyte resistance

The total resistance also can also be calculated with equation (3.39) as:

$$R_T = \frac{\Delta E}{I} \tag{3.39}$$

where:
$\Delta E$ = anode driving potential
$I$ = current density required to achieve cathodic protection

The structure-to-electrolyte resistance ($R_c$) in equation (3.38) can be calculated using equation (3.40) as:

$$R_c = \frac{R}{A} \tag{3.40}$$

where:
R = average coating resistance (ohms/ft$^2$). The supplier specifies R
A = the structure's surface area (ft$^2$)

Assuming $R_W$ in equation (3.38) is negligible, the anode-to-electrolyte resistance can then be calculated from equation (3.41) as:

$$R_a = R_T - R_c \tag{3.41}$$

Equation (3.41) gives the maximum allowable groundbed resistance; this will give the minimum number of anodes required. The number of anodes required is calculated with equation (3.42) as:

$$N = \frac{(0.0052)\rho}{(R_a)(L)}\left[\ln\frac{8L}{d} - 1\right] \tag{3.42}$$

where:
 N = number of anodes
 ρ = the soil resistivity (ohms)
 $R_a$ = the maximum allowable groundbed resistance (*ohms*) as computed in equation (3.41)
 L = the length of the backfill column (ft) (L is specified by the supplier)
 d = the diameter of the backfill column (ft) (specified by the supplier)

5. Calculate the number of anodes for systems life expectancy.
 Each cathodic protection system should be designed to protect a structure for a given number of years. To meet this lifetime requirement, the number of anodes (N) must be calculated using equation (3.43) as:

$$N = \frac{(L)(I)}{49.3(W)} \tag{3.43}$$

where:
 L = design life (years)
 W = weight of one anode (pounds)
 I = the current density required to protect the structure (milliampere)

6. Select the number of anodes to be used. The greater value of equation (3.42) or equation (3.43) will be used as the number of anodes needed for the system.
7. Select the groundbed layout. After calculating the required number of anodes, the area to be protected by each anode is calculated by equation (3.44):

$$A = \frac{A_T}{N} \tag{3.44}$$

where:
 A = the area to be protected by one anode
 AT = the total surface area to be protected
 N = the total number of anodes to be used

For sacrificial anode cathodic protection system, the anodes should be spaced equally along the structure to be protected.

8. Calculate the life-cycle cost for the proposed design. The NACE standard RP-02 should be used to calculate the system life-cycle cost. The design process should be done for several anodes choices to find the one with the minimal life-cycle cost.
9. Organize plans and specification.

## 3.4  GALVANIC ZINC APPLICATION

Zinc and zinc alloys are used as a metal coating, which protects the base material, steel, against corrosion. Zinc in contact with iron (steel) acts as anode—it dissolves completely and covers with a layer of oxides, which inhibits further corrosion. For zinc coatings, pores are not so crucial.

Galvanizing, while using the electrochemical principle of cathodic protection, is not actually cathodic protection. Cathodic protection requires the anode to be separate from the metal surface to be protected, with an ionic connection through the electrolyte and an electron connection through a connecting cable, bolt, or other device. This means that any area of the protected structure within the electrolyte can be protected, whereas in the case of galvanizing, only areas very close to the zinc are protected. Hence, a larger area of bare steel would only be protected around the edges.

Galvanizing protects the underlying iron or steel in the following main ways:

* The zinc coating, when intact, prevents corrosive substances from reaching the underlying steel or iron.
* The zinc protects iron by corroding first. For better results, application of chromates over zinc is also seen as an industrial trend.
* In the event the underlying metal becomes exposed, protection can continue as long as there is zinc close enough to be electrically coupled. After all of the zinc in the immediate area is consumed, localized corrosion of the base metal can occur.

The methods of application available are:

1. Zinc metalizing (plating)
2. Zinc-rich paints
3. Hot-dip galvanizing

### 3.4.1  Zinc Metallizing (Plating)

Zinc metallizing or plating is feeding zinc into a heated gun, where it is melted and sprayed on a structure or part using combustion gases and/or auxiliary compressed air (Langill, 2006).

Zinc metallizing is the thermal spray application of a zinc and aluminum coating to steel. The gas flame bonding of the zinc to the grit-blasted steel is needed to remove any rust or mill scale and prepare the steel for proper adhesion of the zinc. This process protects the steel from corrosion for decades longer than paint alone. This process has been used around the world for 90 years. Steel of every shape and size may be metallized either before or after installation.

Metallizing and hot-dip galvanizing are two zinc-based coatings that protect the steel substrate by a physical barrier and cathodic protection. However, these two coatings are significantly different. Metallizing relies on a mechanical bond between the zinc and the surface of the steel substrate to form a protective coating. Because of this mechanical bond, the surface preparation is critical to performance. Hot-dip galvanizing is a total immersion process where the steel element is dipped into a bath of molten zinc.

### 3.4.2   ZINC-RICH PAINTS

Zinc-rich paints contain various amounts of metallic zinc dust and are applied by brush or spray to properly prepared steel.

Zinc-rich paints are those paints that contain a suitably high amount of zinc dust or zinc powder mixed with organic or inorganic binders. Such zinc-rich paints are applied as a topcoat on steel or other metallic surfaces that operate in harsh environmental conditions and that have a continuous risk of corrosion. The zinc dust prevents the metal from becoming corroded by simply sacrificing itself.

Zinc dust is a zinc material in the form of a powder that is also known as zinc powder.

Zinc-rich paints are primarily used to protect metallic surfaces from corrosion by providing cathodic protection at the expense of the zinc contained in these paints.

Zinc-rich paints and coatings are often used with the addition of a primer that acts as a secondary shield to protect the surface from corroding.

### 3.4.3   HOT-DIP GALVANIZING

Hot-dip galvanization is a form of galvanization. It is the process of coating iron and steel with zinc, which alloys with the surface of the base metal when immersing the metal in a bath of molten zinc at a temperature of around 840°F (449°C). When exposed to the atmosphere, the pure zinc (Zn) reacts with oxygen ($O_2$) to form zinc oxide (ZnO), which further reacts with carbon dioxide ($CO_2$) to form zinc carbonate ($ZnCO_3$), a usually dull gray, fairly strong material that protects the steel underneath from further corrosion in many circumstances. Galvanized steel is widely used in applications where corrosion resistance is needed without the cost of stainless steel and is considered superior in terms of cost and life cycle. It can be identified by the crystallization patterning on the surface.

Features of a hot-dip galvanizing coating are (Langill, 2006):

- Zinc-iron intermetallic layers.
- Harder than the substrate steel.
- Zinc patina.
- Barrier protection.

- Cathodic protection.
- Metallurgical bond to the substrate steel.
- Paintable.
- Edge and corner protection.
- Zinc is a natural and healthy metal.

According to Langill (2006), the benefits of a hot-dip galvanizing coating are:

- No touch-up required
- High- and low-temperature performance
- Application independent of weather
- 100 percent recyclable
- Long-term performance in soils, water, and chemical environments
- Maintenance-free for 50–100 years in most atmospheric environments (Figure 3.39)

Steel articles are inspected after galvanizing to verify conformance to appropriate specifications. Surface defects are easily identified through visual inspection. Coating thickness is verified through magnetic thickness gauge readings.

Zinc Metallizing          Zinc-rich Paints

**FIGURE 3.39** Galvanic zinc applications. (Source: Langill, Thomas J. 2006. *Corrosion protection*. Course notes, Iowa: University of Iowa.)

# 4 Internal Corrosion Protection

Internal corrosion refers to corrosion occurring on the inside of a pipeline. This type of corrosion often results from the presence of molecules such as carbon dioxide ($CO_2$), hydrogen sulfide ($H_2S$), water, organic acids, microorganisms, and other molecules. Typically, these molecules react with the internal pipe surface through anodic and cathodic reactions. The product of these reactions may deposit within the pipe, creating a protective layer that inhibits further corrosion. In other cases, the products do not precipitate and facilitate high rates of corrosion. The rate of internal corrosion depends on the concentration of these corrosive molecules, the temperature, the flow velocity, and the surface material.

For midstream pipelines, a problem called "black powder" occurs because sales gas that is presumed to be dry can still contain some water vapor, which can condense and cause corrosion. The corrosion product can then settle as a powder in the pipeline as well as clog or erode valves and metering equipment. Black powder in gas pipelines is usually corrosion products formed by internal pipeline corrosion but can also be particles from mill scale, weld splatter, formation cuttings, salts, and so on. In order to reduce black powder formation in midstream pipelines, the moisture content should be kept low, and measures to reduce oxygen contamination should be taken. This should include avoiding using air for drying after hydrotesting and eliminating possible sources of oxygen ingress during operation.

Methods of internal corrosion protection used in the oil and gas industry include:

- Internal coatings
- Dehydration
- Chemical injection (e.g., corrosion inhibitor, biocide)
- Cleaning pigging
- Buffering
- Internal cathodic protection (only for internal protection of tanks)

The sections that follow describe the internal corrosion protection methods listed above.

## 4.1 INTERNAL COATINGS

The principal aim for internally coating pipelines is to reduce pipeline friction and internal corrosion. Lower friction and corrosion result in lower operating and installation costs, higher product purity, and increased throughput. It has long been established that a protective coating for pipe cores gives countless savings to pipeline operators. The internal coating provides protection against corrosion and abrasion and it reduces the cost of scrubbers, strainers, "pigs," and other types of pipeline cleaning services. Internal coating ensures product purity, prevents contamination from corrosive products, greatly reduces maintenance and labor costs, provides protection of the pipe interior against the accumulation of deposits (calcareous or paraffin), and as has been well established, substantially increases throughput of product.

For both liquid and gas pipelines, the cost of internally coating the pipeline can be justified in most cases only based on reduced operating costs (Kut, 1975).

Early field testing with gas pipelines showed that, depending on the pipelines and flow characteristics, an increase in throughput of 5–10 percent is possible for 24-inch pipelines (Klohn, 1959). A potential increase of 1 percent can justify the cost of internal coating; this measured increase would appear to give an economic incentive. However, for most applications, contractual supply and/or production considerations rule out the case for the application of internal coating based on either increased product throughput or reduced pipeline costs. Therefore, there is relatively little merit in coating the interior of gas pipelines as a means of capacity enhancement.

The pipeline roughness achievable on the internal surface varies with the internal coating material used, but for maximum improvement in hydraulics, the wall roughness should be around 5–10 microns for gas lines (Singh and Samdal, 1987).

Known benefits associated with internal coating of liquid pipelines include reduced maintenance and lower wax deposition (Jorda, 1966). Up to a 25 percent reduction in wax deposition has been shown to be achievable by application of internal coatings (Singh and Samdal, 1987).

For liquid systems, the economic benefits are greater for smaller pipe diameters, whereas for gas systems the benefits are greater for larger pipe diameters.

## 4.1.1  EPOXY PIPE COATING

Two-pack epoxy-type internal pipe coatings are used in the interior of pipes used in transmitting dehydrated natural gas, wet gas, crude oil, sour crude oil, salt water, drinking water, fresh water, petroleum products, and numerous chemicals. Such specialized epoxy internal pipe coatings have now been available for a considerable number of years, and because of field experience, can be applied in adequate film thickness with the required resistance characteristics.

Two main methods for applying internal coatings are:

1. Spraying
2. In situ coating

## 4.1.2  BENEFITS OF INTERNAL COATING TO GAS PIPELINES

Practical testing and experience have shown that application of internal coating results in the following savings for gas line service (Kut, 1975).

1. Increase in throughput, of 4–8 percent: The increase will be maintained for years, as has been experienced worldwide for many years. However, there is a substantial margin, since it is generally considered that even a 1 percent

improvement in throughput justifies internal coating. For smoothing the internal profile so that gas or fluids flow more readily through the pipe, a thin-film epoxy coating is applied of 1.5–3 mils (37–75 microns) dry film thickness. Application is normally by spraying, following sound surface preparation. This system is essentially used for natural gas pipe insides.

The first major, and now classic, test was in 1958, the "Refugio test." Extensive tests were run at Refugio, Texas, on pipelines operated by the Tennessee Gas Transmission Company. This testing conclusively proved that the Copon internal pipe coating used increased the capacity of the pipe-line—this being the basic proof of increased pipeline capacity, where readings were taken before and after coating.

A 24-inch, 12-mile pipeline was internally coated. After the lines were coated and dry, the increase in flow efficiency was up to 10 percent overall. A 4 percent increase was attributed to cleaning and up to an additional 6 percent to the coating, depending on the rate of flow.

A further test one year later on the same section of pipeline showed there to have been no apparent deterioration of the flow. Further subsequent tests have indicated no significant reduction in the capacity; this has been confirmed by the very substantial savings obtained by initially coating. Furthermore, uncoated pipes require frequent cleaning, unlike internally coated pipes. These results have been closely paralleled by flow results obtained by many transmission companies since 1958.

Klohn (1959) has fully described the testing procedure for establishing this improved throughput. Reference must also be made to a publication by the American Gas Association (1965). This document studies in detail the steady flow in gas pipelines, considering testing, measurement, behavior, and computation. Among other things, this document pertains to internally coated lines. Taylor (1960) has stated that an increase of even only 2 percent in gas flow can justify the cost of internal coatings. The degree of smooth-ness of an internally coated pipe is inversely proportional to the friction resistance. The application of such thick films has been repeatedly shown to provide characteristics that are equal or superior to those of new clean pipes or to regularly pigged pipelines.

2. Easier and faster cleaning of the transmission line after laying, and more rapid drying after hydrostatic testing.
3. Improved flow: In 1846, Jean Louis Poiseuille, and a few years later Darcy, first developed equations for flow, later refined by Reynolds, who conceived "direct" or "streamline flow," and "turbulent flow." Later scientists further developed the concept, leading to the Reynolds Number and recognition of the adverse effect of pipe roughness on flow in a pipe. The Reynolds Number is an index of the amount of "confusion" inside the pipeline. The greater the Reynolds Number and the rougher the pipe surface, the greater the degree of "confusion"; the smoother the pipe surface, the more orderly the flow pattern and the lower the energy loss due to friction. The theory was confirmed by the US Bureau of Mines tests in 1956 and the Refugio test in 1958.

4. Pipe length protected prior to laying: No corrosion which would damage the smoothness and create product contamination.
5. Reduction in paraffin and other deposition, which reduces gas flow.
6. Reduced pumping costs, which are maintained in service.
7. Reduced maintenance: Frequency of cleaning is substantially reduced.

A considerable decrease is achieved in the maintenance of coated lines due to less frequent pigging being required and due to easier cleaning. It was found that in running pigs, about half the pressure was required to move a pig through a coated line than an uncoated line. Similarly, when pipelines were hydrostatically tested it was possible to completely dry a 36-inch pipeline with as few as four pig runs. Frequency of pigging varies from pipeline to pipeline, but several major gas transmission companies have provided some data from their experience. This data showed that with coated pipe, pigging was necessary only every 12–18 months. In uncoated pipe, pigging is normally required about three times a year.

8. Product purity: No contamination from corrosion dust which might block, or damage, applications.
9. Helps pipe inspection: The light-reflecting internal coating highlights lamination and other pipe defects.
10. Reduction in friction:

Published data on loss of pressure in water lines clearly shows that internal coating is not only a vital requirement for protection and maintenance of the installation but that the coating also has a direct effect on losses of pressure and energy.

| Condition | Absolute Roughness ($k_s$ [mm]) |
| --- | --- |
| New, bitumen coated | 0.01–0.02 |
| New, not bitumen coated | 0.04–0.10 |
| Bitumen, partially loosened | 0.08–0.10 |
| Light encrustation | 0.10–0.20 |
| Cleaned after extended use | 0.10–0.20 |
| Overall rusting | 0.15–0.40 |
| Chlorinated rubber coating | 0.007 |
| Two-component polyurethane | 0.001 |

11. Sound economics

In achieving the above benefits, the initial cost of the coating operation is recovered many times. Even if the diameter of the gas pipeline as installed is adequate for the immediate throughput requirements, internally coating is still considered advisable in order to allow a margin for the inevitable increased future demand.

An economic higher compromise figure of 3 mls (75 microns) dry film thickness is specified where the gas is mildly corrosive.

### 4.1.3  BENEFITS OF INTERNAL COATING TO WATER PIPELINES

Because the effect of the condition of the pipe interior is dependent on the Reynolds Number, the effect of internal coatings on the throughput of any line will be greatest for small-diameter lines operating at high capacities. In terms of pressure, the effect of internal coatings will decrease pressure losses in any pipeline more dramatically when the pipeline is smaller and has greater capacity (Kut, 1975).

For example, in a new 6-inch pipeline handling 1000 gpm, internal coating will decrease the pressure drop, compared to the drop in a new, uncoated steel line, by 6 percent; however, in a new 6-inch pipeline handling a mere 300 gpm, internal coating will decrease the pressure drop by only one percent. When the interior of the pipe is extremely rough, as in a five-year-old water pipeline, the effect of the Reynolds Number decreases; and the friction factor is practically constant for all flow rates.

In the instance considered, the improvement in throughput attained by internal coatings will be greater, regardless of the pipeline delivery. Cleaning and coating an old pipeline will decrease pressure drop by as much as 65 percent.

Special linings are available which have been approved by relevant authorities for use with potable water. These coatings have been used in pipelines for potable water and have also given good service in potable water tanks.

1. Reduced pumping cost

    It is generally acknowledged that pumping costs in uncoated lines progressively increase with time. Parallel with the maintenance of throughput efficiency in coated lines, there is no increase in pumping time and costs.

    Seedoriff reports calculations from data taken at the La Huerta pump and at the booster, comparing throughput before and after internally coating a 14-inch diameter plant water pipeline. The total distance of the pipeline was 49,750 ft, with a difference in elevation of 40 ft. Before internal coating, the total friction head was 548.7 ft of water and the pumping rate of 1750 gpm. This compared after internal coating with a pumping rate of 2200 gpm, with a total friction head of 233.9 ft of water.

    Calculations indicate that the cost of power only for pumping 1750 gpm, before internal coating for 8760 hours per year, totaled $367,330. The cost of pumping 2170 gpm after internal coating for 8760 hours per year totaled $350,463. That is a total calculated saving of $16,867 per year in pumping cost alone.

2. Reduction in friction

    In an eight-mile, 16-inch steel water pipeline, coupons were taken from the pipeline after some 11 years to check the internal coating for buildup of calcium carbonate. All the coupons indicated the pipeline to be in practically the same condition as it was when coated, and the C factor remained about 150.

    Thirty-one miles of 10-inch and 12-inch steel water supply pipelines, prior to internally coating, had been in service approximately 11 years. The C factor had dropped from approximately 150 to approximately 65 due to oxygenation barnacles and calcium carbonate buildup. After internally coating, the capacity of the pipeline almost doubled. In 1972, the pipeline was reported as still operating at a C factor approaching 150.

Twenty miles of 16-inch and 8-inch main steel water pipeline and gathering system, when last reported in 1972, had been in continuous service since 1963 with no appreciable decrease in the C factor and no corrosion or evidence of calcium carbonate buildup.

In another typical potable water pipeline, tuberculation had reduced the flow. Internal cleaning and internal coating increased the C factor from 75 to 147, which remained at this figure when rechecked after three years.

Cast iron 1.5-mile, 6-inch pipeline operating at 100 psi, internally coated to improve flow efficiency, was installed some three years earlier but was never cleaned. Throughput had been cut by half, to 213 g at 105 psi. Examination four years after internal coating showed the internal coating to be in excellent condition with a flow of 535 g at 65 psi.

### 4.1.4 SPRAY LINING

Specific spray and cleaning equipment is employed for internally coating small- and large-diameter pipes length-by-length, with appropriately formulated and field-tested epoxy-type coatings, following on sound surface preparation such as abrasive blasting (preferably) or acid cleaning.

Specialized portable equipment is available for the internal blasting of pipe lengths. By inserting a lance fitted with a tungsten carbide deflector, the abrasive stream is deflected radially and strikes the wall of the pipe approximately at right angles. Such simple equipment serves for pipe up to 5-inches in diameter, but for larger-diameter pipes a self-rotating twin jet blast head is normally used. To ensure even cleaning of the pipe wall, the blast head is held centrally and supported by a rolling, adjustable, lightweight carriage (Kut, 1975).

A centering carriage is provided for pipe up to 42-inches in diameter. The spinning blast head and carriage are rolled backward or forward as one unit within the pipe in order to remove mill scale, rust, and chemical contaminants. This method, using inexpensive ancillary equipment, offers a practical and effective means of cleaning larger-diameter pipe. Similar procedures are adopted for internally spraying pipes with the selected pipe coating to the designed film thickness.

### 4.1.5 IN SITU COATING

"In situ coating" or "in place" coating permits the coating of lines already laid— new or old—and avoids the welding problem. The procedure has been used in the United States for some 20 years and for a considerable number of years in Europe and elsewhere.

#### 4.1.5.1 Procedure

One basic procedure consists of two specially designed pup joints fitted with compressed air ports suitably valved, air pressure regulators and bypass units, a coating material inlet port, and a quick-release full diameter and port. Compressed air (or natural gas) is used for driving the specially designed coating plugs.

Two specially designed rubber plugs, comprising rubber discs and rubber bells held together by a mandrel or screwed inserts, contain a batch of coating material between them and are propelled through the pipeline by means of dehumidified compressed air. The fit of the plugs in the pipe is carefully controlled, and the required thin film is deposited behind the rear plug. The thickness is a function of the properties of the coating material, the speed of the coating "train" through the pipe, and the fit of the plugs in the line. The "train" is moved by applying gas or air pressure to the rear plug; the back pressure is controlled by maintaining a known constant pressure differential which controls the speed.

As the paint plugs are propelled by differential positive pressure, the in situ coating is therefore forced into close contact with the pipe wall, flowing into surface irregularities.

Alternatively, in more sophisticated recently developed techniques, the liquids are moved only with synthetic rubber pigs propelled by compressed air; the material is not enclosed between two balls, cups, or pigs (Kut, 1975).

### 4.1.5.2   In Situ Surface Preparation
Surface preparation by abrasive blasting is not employed. It is generally carried out by first removing loose scale and rust by passing brushes and scrapers with water and detergent through the line, followed by special techniques of acid cleaning.

Tar, grease, or other contaminants are removed with solvent or, if necessary, emulsion flushes, and passed as frequently as required, in combination with brushes and scrapers. Variations in the process by applicators include neutralization with ammonium hydroxide after each acid cleaning stage.

In pickling the line with inhibited hydrochloric acid, the concentrations employed and time of surface contact are dependent on the pipe interior condition. Acid residue must be flushed with fresh water, and the surface is then usually lightly phosphate-washed acid free, the water removed, and the pipe finally dried, preferably with dehumidified air.

### 4.1.5.3   In Situ Lining
The internal pipe coating is then applied in a similar manner—calculated excess volume is introduced and propelled by dehumidified compressed air. To ensure good wetting and coverage, the coating is reversed. In a multicoat application, adequate drying time is allowed between coats, depending on the prevailing ambient conditions. Solvent removal is important to ensure freedom from holidays, with a dry film thickness varying from 250 to 400 microns. The film thickness is closely monitored. The design, fit, construction, speed, and number of passes of the plugs and coating trains to achieve the required film thickness is very much a matter of the expertise and experience of the in situ coating contractor (Kut, 1975).

### 4.1.5.4   Pipeline Design for In Situ Coating
The length of pipe that can be coated and cleaned in one operation depends on the configuration of the pipeline and averages 3–5 miles, though a larger continuous run is possible.

The pipeline should have no bends sharper than 5D radius, although in some instances 3D bends can be negotiated and should have no branch-offs or reducers. The ends should be in a horizontal plane and flanged to permit attaching the inlet and receiving traps, for which a space of some two to three meters is required. Excessive weld penetration in the form of weld icicles should be avoided in order to allow for the passage of a 98 percent internal diameter gauging pig. Tees or valves have to be replaced with temporary flanged spools, which at the same time serve as inspection points. Incorporation of one or more inspection spools may be desirable.

Film thickness is controlled by the speed of travel and pressures as well as the excess of calculated volume of coating charged into the line.

Another in situ coating procedure involves the use of a train treating relatively shorter lengths at a time of steel or concrete pipelines. Mechanical cleaning is with hammers and water flushing, or with abrasive blasting. A specially developed head within the "train" applies the coating by brush or spraying. At present, there are certain limitations in the smaller diameters which can be coated. A traveling TV camera inspects the work (Kut, 1975).

### 4.1.5.5 Testing In Situ Coating

Sophisticated examination has been periodically suggested, such as with TV cameras. However, trials have shown these instruments to be too fragile to stand up to travel within a pipeline, particularly if there is—due to a bend or high spot—a stoppage and then release with a sudden surge. Another problem is location of any defect shown by the camera. Reliance is therefore placed on test spools and simple mirrors. By employing mirrors to reflect sunlight on inaccessible portions of the pipeline, a clear view of the surface condition may be obtained for distances of up to 20 meters in pipe of up to 10 inches in diameter and up to 40 meters in pipes in excess of 10 inches in diameter. Similar results may be obtained with spotlights. Practical experience confirms the effectiveness of these procedures (Kut, 1975).

The Solvent Rub Test is usually performed using methyl ethyl ketone (MEK) as the solvent. Solvents, specifically MEK, is used to measure the level of cure of coating systems. ASTM D4752 involves rubbing the surface of a baked film with cheesecloth soaked with MEK until failure or breakthrough of the film occurs. The type of cheesecloth, the stroke distance, the stroke rate, and approximate applied pressure of the rub are specified. The rubs are counted as a double rub (one rub forward and one rub backward constitutes a double rub). The solvent rub test provides a quick relative estimation of degree of cure without having to wait for long-term exposure results. However, the method's test results can be misleading, making it difficult to quantify cure response.

### 4.1.6 TREATMENT OF WELD

One of the major problems in internally coating pipes length-by-length before laying is subsequent treatment of the weld, where this is essential and couplings cannot be used.

In large-diameter pipes, the welds can be made manually good, though surface preparation is likely to be of lower order. However, this is not possible with smaller diameters.

The remote-controlled auto-detection machine engages the pipe surface at each joint on command and brushes the area of burned coating caused by welding. On completion of the cycle, the machine automatically cuts off. After a vacuum pipe-cleaning tool is run through the pipe to collect dust and slag, a centrifugal impeller replaces the brushes' head, and the tool is re-run through the pipeline for coating operation. During rotation, the extruded brushes clean the weld as well as both sides. When not rotating, the brushes remain retracted, to avoid damaging the internal pipe coating during travel to and from the work area. To ensure thorough cleaning, the work speed of the carriage is set at about one inch per second. A centrifugal impeller applies the coating. To ensure a constant volume of paint during application, the speed of travel of the carriage and the paint flow are synchronized. The total volume depends on the pipe diameter.

## 4.2 CHEMICAL INJECTION

In the general oil and gas industry, the following chemical additives are stored and added to the process to ensure reliable and safe operation.

### 4.2.1 CORROSION INHIBITOR

These chemicals prevent the corrosion of pipes, pipelines, and tanks. A corrosion inhibitor works by forming a passivation layer on the metal that prevents access of the corrosive substance to the metal.

A corrosion inhibitor reduces the corrosion rate of a metal exposed to that environment. Inhibition is used internally with carbon steel pipes and vessels as an economic corrosion control alternative to stainless steels and alloys, coatings, or nonmetallic composites, and can often be implemented without disrupting a process. Corrosion inhibitors or other corrosion protection chemicals such as mono-ethylene glycol (MEG) are injected into the system.

The major industries using corrosion inhibitors are oil and gas exploration and production, petroleum refining, chemical manufacturing, heavy manufacturing, water treatment, and the product additive industries (NACE 2016).

#### 4.2.1.1 Types of Corrosion Inhibitors

There are four types of corrosion inhibitors:

1. Anodic Inhibitors
2. Cathodic Inhibitors
3. Mixed Inhibitors
4. Volatile Corrosion Inhibitors

##### 4.2.1.1.1 Anodic Inhibitors

An anodic inhibitor acts by forming a protective oxide film on the surface of the metal. It causes a large anodic shift that forces the metallic surface into the passivation region, which reduces the corrosion potential of the material. This entire procedure is sometimes called passivation (Zavenir, 2018). Some examples are chromates, nitrates, molybdate, and tungstate.

Anodic inhibitors are considered dangerous because of their chemical characteristics.

### 4.2.1.1.2   Cathodic Inhibitors

A cathodic inhibitor slows down the cathodic reaction to limit the diffusion of reducing species to the metal surface. Cathodic poison and oxygen scavengers are examples of this type of inhibitor.

Cathodic inhibitors work by two different methods:

1. It may slow down the cathodic reaction itself, or
2. It may selectively be precipitating on cathodic regions to restrict the diffusion of eroding elements to the metal surface.

The cathodic reaction rate can be decreased by the use of cathodic poisons. However, it can also enhance the sensitivity of a metal to hydrogen-induced cracking because during aqueous corrosion or cathodic charging the hydrogen can also be absorbed by the metal (Zavenir, 2018).

The use of oxygen scavengers that react with dissolved oxygen can also decrease the corrosion rates. Examples of cathodic inhibitors include:

- Sulfite and bi-sulfite ions that form sulfates when reacting with oxygen.
- Catalyzed redox reaction by either cobalt or nickel.

### 4.2.1.1.3   Mixed Inhibitors

Mixed inhibitors are film-forming compounds that reduce both the cathodic and anodic reactions. The most commonly used mixed inhibitors are silicates and phosphates used in domestic water softeners to prevent the formation of rust water (Zavenir, 2018).

Mixed inhibitors are film-forming compounds that reduce both the cathodic and anodic reactions. The film-forming solution causes the formation of precipitates on the metal exterior, preventing both anodic and cathodic sides indirectly. Examples of mixed inhibitors include:

- Silicates and phosphates used in residential water softeners to limit the development of rust water.
- In aerated hot-water systems, sodium silicate protects steel, copper, and brass.

### 4.2.1.1.4   Volatile Corrosion Inhibitors

Volatile corrosion inhibitors (VCI), also called vapor phase inhibitors (VPI), are products moved in a closed atmosphere to the section of corrosion by volatilization from a source. For example, in boilers, volatile compounds such as morpholine or hydrazine are transported with steam to prevent corrosion in condenser tubes by counterbalancing acidic carbon dioxide or by changing exterior pH toward less acidic and corrosive rates (Zavenir, 2018).

In closed confined spaces, such as shipping containers, VCI products such as VCI paper, VCI bags, or VCI rust removers are used. When these VCIs come in contact with the metal surface, the vapor of these products is hydrolyzed by any moisture to release protective ions.

It is very important for an efficient VCI to produce inhibition quickly, while lasting for a prolonged period.

The qualities of a VCI product depend on the volatility of its compounds. Quick-action sequence high volatility, while providing protection, requires low volatility. Examples of volatile corrosion inhibitors include:

- In boilers, volatile mixtures such as morpholine or hydrazine are carried with steam to stop corrosion in condenser pipes.

### 4.2.1.2 Applications of Corrosion Inhibitors

- Volatile amines are used in boilers to minimize the effects of acid. In some cases, the amines form a protective film on the steel surface and at the same time act as an anodic inhibitor. An inhibitor that acts both in a cathodic and anodic manner is termed a *mixed inhibitor*.
- Benzotriazole inhibits the corrosion and staining of copper surfaces.
- Corrosion inhibitors are often added to paints. A pigment with anticorrosive properties is zinc phosphate. Compounds derived from tannic acid or zinc salts of organonitrogens (e.g., Alcophor 827) can be used together with anticorrosive pigments. Other corrosion inhibitors are Anticor 70, Albaex, Ferrophos, and Molywhite MZAP.
- Antiseptics are used to counter microbial corrosion.
- Benzalkonium chloride is commonly used in the oil field industry.
- In oil refineries, hydrogen sulfide can corrode steels, so it is removed often—using air and amines by conversion to polysulfides.

### 4.2.2 Scale Inhibitor

A scale is a deposit of insoluble inorganic mineral. Common oilfield scales include calcium carbonate and barium sulfate. Scale deposition within processing units such as pipes and heat exchangers obstructs or blocks fluid flow. A scale inhibitor inhibits scale formation and deposition.

Scale-inhibiting chemicals that are applied up or downhole of the wellhead and are, in general, classified into four categories:

- Oil-miscible
- Totally water-free
- Emulsified
- Solid

Subject to the mineral content present in the water, duration of the job, and operational needs, the chemical(s) can be applied continuously or in scale-squeeze applications.

### 4.2.3  HYDRATE INHIBITORS

Hydrate formation along a long natural gas pipeline has recently been established to initiate different types of internal corrosion along the pipe length based on the formation stage and point. These corrosions may lead to disintegration of the pipe's properties and eventually result in the pipe's leakage or full-bore rupture. Apart from the enormous economic implications on the operating company, the conveyed fluid, upon escape to the environment, poses the risk of fire, reduction of air quality, and other health hazards.

Hydrate inhibitors are used to inhibit hydrate formation (e.g., methanol). Formation of hydrates may give rise to operational problems such as heat exchanger tube blocking, instrumentation plugging, or pipeline blocking and internal corrosion; hence, the conditions suitable for their formation should be avoided whenever possible.

The conditions at which hydrate formation is most likely to occur correspond to high pressure and low temperature. For pipeline systems, the lowest temperature normally occurs during shutdown, so this situation must always be considered. Free water must be present within the pipeline or processing system. Further conditions that are known to promote hydrate formation are:

- High velocities
- Pressure pulsations
- Agitation

The hydrate crystals forming on pipe walls initiate at any unevenness such as weld seams or areas of corrosion. Narrowing of the internal pipe diameter starts slowly, but as it changes flow and pressure conditions, hydrate buildup is accelerated.

Pipelines that are operated in the presence of free water at hydrate-forming conditions must be injected continuously with a hydrate point depressant or hydrate inhibitor. The most common hydrate inhibitors used are methanol, ethylene glycol, or diethylene glycol.

At pipeline conditions below minus 10°C, glycols are not often used due to their high viscosity. Above minus 10°C, glycols may be preferred due to their lower injection rate requirement. The required injection rate of glycols is generally less than for methanol due to the lower vaporization rate of the glycols into the gas. The concentration of hydrate point depressant or hydrate inhibitor required in the liquid phase to prevent hydrate formation is given by the following equation:

$$W = \frac{dM}{k + dM} \times 100 \qquad\qquad (4.1)$$

where:
- W = weight % inhibitor (depressant) in the liquid phase
- d = depression of hydrate point required (°C)
- M = molecular weight of depressant or inhibitor (32 for methanol, 62 for ethylene glycol, and 106 for diethylene glycol)
- k = constant (1297 for methanol, 2220 for glycols)

The rate of inhibitor addition required in the liquid phase can be evaluated, knowing the quantity of free water in the system.

For pipelines that normally transport dry gas, the most likely occasion at which hydrate formation may pose a problem is during commissioning. Prior to commissioning with hydrocarbon gas, the pipeline will have been filled with water in order to carry out a system hydrotest. Hence, between the hydrotest and gassing-up operations, free water must be removed or treated to prevent the formation of hydrates. Numerous methods of drying or treating the pipeline available include:

- Methanol swabbing
- Vacuum drying
- Hot air drying
- Inert gas drying

For long pipelines, methanol swabbing offers the most economical solution. This procedure leaves a film of methanol and water on the pipe wall that has a suitable concentration of methanol to inhibit hydrate formation when gassing up. The other drying procedures prevent the possibility of hydrates forming by completely vaporizing all free water in the pipeline.

The final connection to the pipeline system usually is comprised of a double block valve and bleed system. During the pre-commissioning operations, both of these block valves will usually be closed, leaving the spool piece between the valves full of water. In order to prevent the possibility of hydrate formation during gassing up, this water must be treated. The simplest solution is to fill the spool piece with either methanol or glycol to a suitable concentration that will inhibit hydrate formation at the pipeline operating conditions. Ideally, as much water as possible should be displaced by the methanol or glycol, as this will further minimize corrosion problems that may occur if acid gases dissolve in the free water.

The existence of hydrocarbon hydrates in a system may be detected by a change in the pressure profile within the system. Once hydrocarbon hydrates have started to form, the buildup accelerates. Hence, the system pressure profile will continue to change, giving rise to potential capacity limitations. The removal of hydrocarbon hydrates from a pipeline system involves running a batch of either methanol or glycol through the pipeline, driven by a pig. The size of the batch should be sufficient to depress the hydrate formation temperatures of the maximum anticipated hydrate deposition to below the pipeline temperature as well as allowing for liquid that will be left on the pipe wall and not pushed forward by the pig. Removal of the hydrates from raw natural gas pipelines, which may not have pigging facilities, is usually achieved by increasing the dosage rate of hydrate point depressant. For a short period, shock dosing, at a rate of approximately five times the normal injection rate, should be undertaken. An alternative approach to removing hydrate is to operate the pipeline at different physical conditions. By lowering the operating pressure or increasing the operating temperature, the physical conditions suitable for hydrate formation may be avoided. This solution may be of limited use, as the flexibility of the operating conditions, consistent with the required production rate, are unlikely to be high.

### 4.2.3.1  Hydrate Formation and Inhibition

In production and processing of fluids containing hydrocarbons and water, there is a risk of hydrate formation. Natural gas hydrates are ice-like crystalline compounds formed when natural gas components such as methane, ethane, propane, isobutane, hydrogen sulfide, carbon dioxide, and nitrogen are entrapped in a crystal lattice of water molecules. The water cage-like structure is called a "host" and the entrapped gas molecules are called "guests" (see Figure 4.1).

Their accumulation in processing facilities or pipelines restricts or stops the flow of fluids, causing shutdowns and even destruction of valuable equipment (see Figure 4.2).

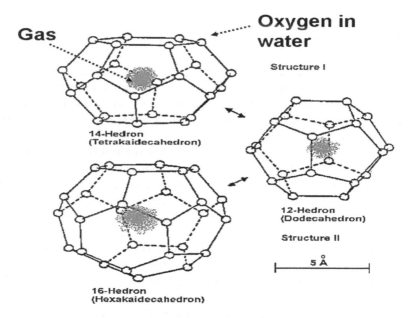

**FIGURE 4.1**  Structures gas hydrates. (Source: Courtesy of Oak Ridge National Laboratory, U.S. Dept. of Energy.)

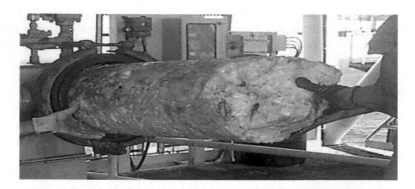

**FIGURE 4.2**  Hydrates blocking pipeline.

### 4.2.3.2  Conditions Necessary for Hydrate Formation

1. Low temperature
   - Hydrates form at low temperatures called formation temperatures. If a gas containing water is cooled below its hydrate formation temperature, hydrates will form.
2. High pressure
   - High pressure favors hydrate formation.
3. Water
   - Free water or water vapor at sufficiently low temperature or high pressure enhances hydrate formation.

Further conditions that are known to promote hydrate formation are:

- High velocities
- Pressure pulsations
- Agitation

It is important to note that:

- High BTU gas is more likely to produce hydrates and freezing problems.
- Due to the Joule-Thomson effect; whenever gas pressure is reduced, its temperature is also reduced. This may result in freezing of the pipeline or formation of ice in the pipeline. Temperature will decrease approximately 7 degrees Fahrenheit for every 100-psi pressure reduction. Temperature will decrease approximately 7 degrees Fahrenheit for every 100-psi pressure reduction.

### 4.2.3.3  Types of Hydrates

Three types of gas hydrates have been identified:

- Type I: The guest molecules are small gas molecules (methane, ethane, $H_2S$, $CO_2$).
- Type II: The guest molecules are larger gas molecules (propane and isobutane and n-butane).
- Type III: The guest molecules include benzene, cyclopentane, and cyclohexane.

### 4.2.3.4  Methods of Hydrate Inhibition

To prevent the formation of hydrates within the process, the following methods are used:

- The temperature is maintained below the hydrate formation temperature of gas.
- The pressure is maintained below the value necessary for hydrates to form.
- Chemicals called hydrate inhibitors (methanol and/or one of the glycols) are injected into the process where hydrates are likely to form. They push

the hydrate formation equilibrium to lower temperature or higher pressures. Certain inhibitors called low-dosage inhibitors keep any hydrate formed from coagulating and growing, or delay the hydrate formation process.
*   Water is removed from the gas stream through the process called gas dehydration.

We shall look at hydrate inhibition methods and gas dehydration next.

### 4.2.3.5   Hydrate Inhibition

Methanol, or one of the glycols, mixes with the condensed aqueous phase to lower its hydrate formation temperature at the given pressure (or increase the hydrate formation pressure at a given temperature) when injected into a gas process stream. Because they shift the equilibrium to lower temperatures and higher pressures, they are termed thermodynamic inhibitors. They also lower the freezing point of the liquid water (see Figure 4.3). To be effective, the chemical inhibitor is injected at the very points where the wet gas is cooled to its hydrate temperature. The injection must be in such a way that there is good mixing with gas stream. Both the glycols and the methanol can be recovered together with the aqueous phase, and then regenerated and recycled. The choice between inhibitors depends on operating conditions and economics.

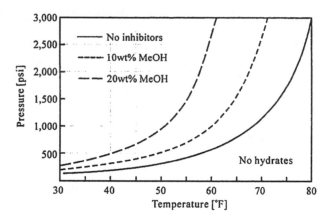

**FIGURE 4.3**   Effects of methanol on hydrate formation. (Source: Subsea pipeline and risers by Yong B. and Qiang B. 2005.)

### 4.2.3.5.1   Methanol Injection

Methanol has a low freezing point. It can be injected at any temperature, and it is very effective as a hydrate inhibitor due to its ability to achieve high dew point suppression. Methanol is preferred for use at low-temperature conditions when there is separation equipment downstream, because glycol is harder to separate at low temperatures (due to its increased viscosity).

Higher injection rates of methanol are required, since some is lost to the hydrocarbon gas phase (due to its higher volatility), and some is lost to the liquid hydrocarbon phase (due to its solubility in liquid hydrocarbons). It is only the methanol that is dissolved in water that inhibits hydrate formation.

The efficiency of inhibition depends on the concentration of methanol injected. It can be injected intermittently or continuously.

*4.2.3.5.1.1   Problems with Methanol Injection*   Methanol can dissolve alcohol-based corrosion inhibitors injected into the process to prevent corrosion leading to unexpected corrosion problems.

Methanol can concentrate in a liquefied petroleum gas (LPG) stream. LPG is made largely of propane and mixed butanes. Methanol forms azeotropes with the propane and butane components in the LPG, and these cannot be separated by distillation.

*4.2.3.5.1.2   Methanol Recovery*   Recovering methanol from the process by distillation is not economical, so in most instances it is not recovered after use. In the gas plant, methanol is not recovered.

### 4.2.3.5.2   Glycol Injection

Ethylene glycol (EG) is the most used of all the glycols in hydrate inhibition due to its low cost, lower viscosity, and lower solubility in liquid hydrocarbons. Losses are generally very small so do not need to be considered when calculating injection rates (see Figure 4.4).

*4.2.3.5.2.1   Glycol Recovery and Regeneration*   Glycol is recovered and regenerated. After injection into the process, the glycol-water solution and the liquid hydrocarbons can form an emulsion during agitation or during expansion from a high pressure to a lower pressure, such as across a throttling valve. To recover the glycol, glycol and the condensed water mixture is separated from the gas stream in a separator. The recovered glycol-water mixture is taken to the glycol regeneration unit where the water is stripped from the mixture and regenerated glycol can then be reused.

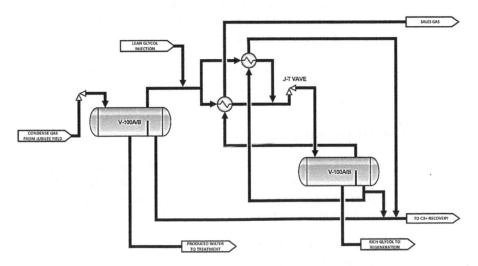

**FIGURE 4.4**   Glycol injection in the Gas process plant.

The regeneration process is designed to produce a glycol solution that must have a freezing point below the minimum temperature encountered in the system if it is to be injected. The freezing point of a glycol solution in water is dependent on the weight percent of glycol in the solution. This is typically 75–80 wt. percent.

Glycol injection also aids in dehydrating the gas.

As noted earlier, the viscosity of glycol increases as temperature decreases. The design of units containing chillers and refrigeration units where glycol is injected need to consider this. If the temperature is too low, the rich glycol solution leaving the unit is very viscous, making downstream separation difficult.

## 4.2.4 BIOCIDES

Biocides prevent microbiological activity in oil production systems. Uncontrolled bacteria, algae, and fungus activities are problematic in oilfield operations. For example, bacteria activity such that of sulfate-reducing bacteria results in the production of $H_2S$, which leads to reservoir souring, metal corrosion, health hazards, and clogged filters. Typical uses include diesel tanks, produced water (after hydrocyclones), slop, and ballast tanks.

Biocides are utilized to inhibit and eliminate microbiologically influenced corrosion (MIC) caused by the corrosive action of microbes. Biocide is injected into the pipeline in the stream of a non-electrolytic carrier. In many cases, the biocide is added to the buffering agent so that only one addition to the gas stream is needed.

Other active agents such as film-formers that aid in forming a passive barrier at the pipe surface and agents that promote the evaporation of electrolytes can also be added. Many of these agents are expensive, and depending on the gas flow at the time of injection or use, may or may not reach the location where the electrolytes and microbes are trapped (Baker, 2008).

Biocides discussed here are used in many industries such as the oil and gas industry. Microbial control in the oil and gas industry is primarily practiced to prevent the detrimental effects of microbial growth on production equipment, pipelines, and the reservoir. There are four main groups of biocides:

1. Preservatives (e.g., biocides for liquid cooling and processing systems, metal-working fluids biocides, and biocides for the oil and gas industry)
2. Disinfectants (e.g., drinking water disinfectants)
3. Pest control
4. Other biocidal products

Antimicrobial agents and corrosion inhibitors are widely used in the oil and gas industry. Treatment chemicals are used in the natural gas industry from well development through transmission and storage of natural gas (Turkiewicz et al., 2013).

Based on their chemical action, biocides can be divided into two groups:

1. Substances with oxidizing effect
2. Substances with nonoxidizing effect

The most commonly used oxidizing biocides are chlorine, bromine, ozone, and hydrogen peroxide. Nonetheless, the use of oxidizing biocides is accompanied with these negative effects:

- Interaction with other chemicals (corrosion inhibitors)
- The possibility of interaction with nonmetallic substances
- Initiation of corrosion of structural materials

Before each treatment with oxidizing preparations, these effects should be taken into consideration when considering the potential for oxidation, the dose, and type of treatment (intermittent or constant).

The group of nonoxidizing biocides includes aldehydes (e.g., formaldehyde, glutaraldehyde), acrolein, quaternary ammonium compounds, amines and diamines, and isothiazolones.

Often used in the industry, quaternary ammonium compounds are used as cationic corrosion inhibitors and biocides. The biocidal activity of these substances is to dissolve the lipid cell membrane, which leads to loss of the cell contents of the microorganism.

*Quaternary ammonium cations, also known as QUATS*, are positively charged polyatomic ions of the structure $NR_4^+$, R being an alkyl group or an aryl group. QUATS prevent the formation of polysaccharide secretions during bacterial colonization, thus slowing antibacterial activity. QUATS are used in closed systems and gas manifolds. However, they are not used during the exploitation of oil because they may adversely affect the permeability of the crude oil deposit. Furthermore, they are not compatible with oxidizing agents, especially the chlorates, peroxides, chromates, or permanganates. Most of these compounds are readily biodegradable. Benzalkonium chloride is a common type of QUATS salt used as a biocide, a cationic surfactant, and as a phase transfer agent.

*Isothiazolones* is another type of biocide. They are fast-acting biocides inhibiting growth, metabolism, and biofilm formation by algae and bacteria. They are used in combination with other biocides or individually, typically aqueous solutions of chloride- and methyl-derivatives of these compounds. Isothiazolones are used only in an alkaline medium. At pH < 7, they lose biocidal properties. Moreover, these compounds can be used in combination with other chemicals without changes in performance (Turkiewicz et al., 2013). An exception is an environment containing hydrogen sulfide, which causes deactivation of isothiazolones. The main application areas of isothiazolones are coolants and cooling and lubrication fluids.

Another compound is *glutaraldehyde* (pentane-1,5-dial). This is the most common component of commercial biocides, with powerful antibacterial and antifungal activity. An important advantage of this compound is the possibility of use in a wide range of temperatures and pH as well as solubility in water. Glutaraldehyde does not react with strong acids and alkalis but reacts violently with ammonia and amine-containing substances, which causes an exothermic polymerization reaction of an aldehyde, and thus its deactivation. It is not sensitive to the presence of sulfides and tolerates high-salinity environments (Turkiewicz et al., 2013).

Another biocide used in industry is *Tetrakis (hydroxymethyl) phosphonium sulfate (THPS)*. It is a water-soluble ionic biocide that destroys bacteria, fungi, and algae in industrial cooling installations and process water tanks. It is characterized by low toxicity and interacts with other chemicals used in aqueous environments. A particular advantage of this compound is its ability to remove residual iron sulfide in pipelines. THPS biocide is a kind of environment-friendly water treatment microbiocide that is made of Tetrakis (hydroxymethyl) phosphonium sulfate (THPS 75 percent) solution. It can withhold sulfate-reducing bacteria (SRB) and most aerobic bacteria, including microorganisms that form biofilm in enhanced oil-recovery process, production, and other supporting systems such as water injection equipment, well water disposal facilities, water-holding tanks, recirculating water treatment systems, and pipelines. THPS biocide is also effective in controlling microbial growth in drilling muds and stimulation fluids for oil and gas wells. Tetrakis (hydroxymethyl) phosphonium sulfate is characterized by its low solidity point and good stability. THPS 75 percent solution can easily dissolve in water and can be preserved for a long time (Turkiewicz et al., 2013).

Structural formula:

In gas storage systems, the traditional treatment is with biocide and/or corrosion inhibitor. The chemical is mixed with water or a suitable solvent, pumped into the storage well, and followed with gas. This method assumes the gas will displace the chemical mixture into the formation to control bacterial growth. Although negative bacteria cultures have been noted with this type of treatment, there is concern that the biocide stayed in the bottom of the well or near the well bore and was not carried far into the formation where bacterial growth still continues.

Side effects of biocide treatments are foaming, emulsions, and expensive equipment costs for individual well treatments.

## 4.2.5 ANTIFOAM

Antifoam is a chemical additive mixed with industrial process liquids so that the formation of foam in the industrial process liquids can be avoided. Some of the commonly used defoamers or antifoaming agents are insoluble oils, glycols, polydimethylsiloxanes, silicones, octyl alcohol, aluminum stearate, and sulfonated hydrocarbons (Corrosionpedia, 2017).

Antifoams prevent the formation of foam in oil processing by reducing liquid surface tension. They are used especially in units such as separators, where foaming impedes the effective separation of gas from the liquids. Antifoam is used to eliminate foam that may cause overflowing, clogging, corrosion, or electrical shorts. If foam is produced in any industrial process liquid, it can cause serious problems such as defects in the surface coatings due to the formation of air bubbles in the surface coatings. Because of this, unevenness occurs, and a smooth surface coating is not achieved.

## 4.2.6  DRAG REDUCERS

Drag reducers, also known as flow improvers of the long-chain polymer type, have been known since 1946, when B. A. Toms, a British chemist in London, first undertook experiments. He found the drag reduction phenomenon while studying the characteristics of liquid solutions in turbulent flows.

Energy must be applied to the fluid being transported through a pipeline. The energy moves the fluid but is lost in the form of friction as the fluid moves down the pipeline.

Flow-improver technology can reduce the energy lost due to friction or drag by more than 50 percent in most cases and increase flow rates by up to 100 percent. The performance depends largely on the properties of the fluid being transported and the condition of the pipeline.

Erosion can occur because of the contact between the pipe wall and the flow of gas or hydrocarbon liquid. The most prominent factors contributing to the levels of erosion include flow rate and the level of contamination present in the gas or fluid. For example, the greater the volume of contamination within the pipeline, the greater the risk of erosion. Additives such as drag reducers/flow improver have the ability to prevent flow-induced localized corrosion (FILC) or erosion corrosion, internal corrosion, and corrosion of the metal wall. Drag-reducing additives reduce freak energy densities to values significantly below fracture energies of protective layers and hence inhibit initiation of FILC (Schmitt and Bakalli, 2008).

### 4.2.6.1  Drag Reduction

Drag reducers improve flow in pipelines. During fluid flow, the portion of the fluid in contact with the wall is slower than the portion in the center. This leads to the formation of turbulent eddies which cause drag. A drag reducer suppresses the formation of turbulent eddies, reducing the drag and subsequently the frictional pressure drop in the pipeline.

Drag is a term that refers to the frictional pressure loss per length of pipe that develops when a fluid flows in a pipeline. Drag increases with increasing flow velocity. Drag reduction is the proportional decrease in this frictional pressure drop achieved with the addition of very small amounts of a specialty chemical that acts as a drag-reducing agent, also called a drag-reducing additive or a flow improver (Dean, 1984).

Drag reduction, as defined by Savins (Dean, 1984), is the increase in pumpability of a fluid caused by the addition of small amounts of an additive to the fluid.

The effectiveness of a drag reducer is normally expressed in terms of percent drag reduction.

At a given flow rate, percent drag reduction is defined as:

$$\% D.R = \frac{P - P_p}{P} \times 100 \qquad (4.2)$$

where:

   P = base pressure drop of the untreated fluid
   $P_p$ = pressure drop of the fluid containing drag-reducing polymer
   D.R = Drag Reduction

Generally, the increased pumpability is used to increase the flow rate without exceeding the safe pressure limits within the flow system. The relationship between percent drag reduction and percent throughput increases can be calculated using the following equation:

$$\text{Percent throughput increase} = \left[ \left( \frac{1}{1 - \dfrac{\% D.R}{100}} \right)^{0.55} - 1 \right] \times 100 \qquad (4.3)$$

where:

$\% D.R =$ is the percent drag reduction as defined in equation (4.2).

Equation (4.2) assumes that pressure drop for both the treated and untreated fluid is proportional to the flow rate raised to the 1.8 power.

In most petroleum pipelines, the flow through the pipeline is turbulent. Turbulent flow is characterized by irregular, random motion of fluid particles in directions transverse to the direction of the main flow. The flow is unstable. Turbulent eddies are generated at the pipe wall and move into the core of the pipe. More energy is required to transport fluid at a given average flow velocity in turbulent flow because not all of the energy is dissipated in the formation of eddy currents (Figure 4.5).

**FIGURE 4.5**   Flow regimes in a pipeline. (Source: Dean, Hale. June 1984. "Special Report-Slick Way to Increase Capacity." *Pipeline & Gas Journal* 17–19.)

In most cases, a general family of polymeric chemical additives called drag reducers can decrease this turbulent energy loss. Generally, the more turbulent the flow, the more effective the drag reducer becomes and consequently, the more efficient energy utilization can be achieved (Figure 4.6).

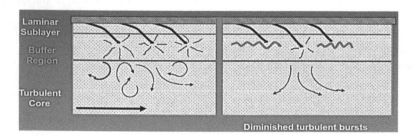

**FIGURE 4.6**   Drag reduction mechanism. (Source: Laura Thomas, Tim Burden. 2010. *Heavy oil drag reducing agent (DRA): increasing pipeline deliveries of heavy crude oil.* UK: ConocoPhillips Specialty Products Inc.)

Turbulence is first formed in the buffer zone, and drag reducers are most active in the buffer zone. Actual performance depends on the hydraulic characteristics of the pipeline and the physical properties of the liquid.

### 4.2.6.1.1   Pipeline Drag Reducing Additive (PDRA)

A pipeline drag-reducing additive is typically a high-molecular-weight hydrocarbon polymer suspended in a di-hydrocarbon solvent. When mixed with a fluid such as crude oil or refined petroleum products in pipelines, it changes the flow characteristics and reduces the flow turbulence in the pipeline.

The strength of the turbulent eddy currents at the pipe wall are reduced by the addition of a drag reducer. Some believe that PDRA absorbs part of the turbulent energy and returns it to the flowing stream. By lowering the energy loss or drag, the PDRA allows the pipeline throughput to increase for a given working pressure, thereby increasing normal pipe capacity or throughput or to operate at a lower pressure for the same throughput, thereby decreasing operating cost.

Pipeline drag-reducing additives do not work by being absorbed into or coating pipelines but rather but by being dissolved into and becoming part of the fluid. The PDRA is highly susceptible to shear stresses in a pipeline system and thus loses its effectiveness readily. This happens mainly as the internal shear stresses of the pipeline flow and geometry break the PDRA long-chain molecules into smaller pieces. Thus, the PDRA must be added continuously to the pipeline fluid to maintain the desired level of drag reduction.

### 4.2.6.1.2   Injection of Flow Improvers

A compact, skid-mounted chemical injection module consisting of injection pumps, positive displacement flowmeters with totalizers, and miscellaneous instrumentation is used to inject the polymer. The polymer is viscous, so an inert gas, usually nitrogen, is used to help transfer material from tanks to pumps (Figure 4.7).

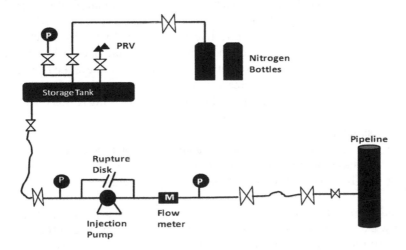

**FIGURE 4.7**   Injection system for drag reducer/flow improver. Source: Dean, Hale. June 1984. "Special Report-Slick Way to Increase Capacity." *Pipeline & Gas Journal* 17–19.

### 4.2.6.1.3   Economics of Flow Improvers
The economics of using a drag reduction additive depends on:

- How much drag reduction or flow increase is needed
- The characteristics of the crude oil/petroleum products being transported
- The pipeline configuration

### 4.2.6.1.4   Limitations
For the drag-reducing additives to work:

- The flow through the pipeline must be turbulent.
- The additives are usually effective up to more than 70 percent flow increase.
- For flow increase above 50 percent, a combination of new pumps or loop sections together with additives may be required.

### 4.2.6.1.5   Applications of Flow Improvers
Initially it was thought that only very-large-diameter lines delivering huge volumes could benefit from use of drag reducers. Tests have demonstrated, however, that smaller crude and product lines also can benefit. The use of flow improvers in pipelines results in the following advantages:

1. Flow increase
   - Increased flow rate of one or many pipeline segments
   - Increased tanker loading rate
   - Increased floating storage offloading (FSO) loading rate

The most common use of drag-reducing additives is to increase pipeline flow rates in a system that is at capacity. In order to deliver more hydrocarbon fluids in the system, DRA is injected at each segment to increase the overall pipeline capacity.

2. Reduces capital expenditure
      DRA usage is very cost-effective versus capital-intensive mechanical expansion (limits the need for additional pumping facilities or looped pipe segments). In addition to no capital requirements to achieve the extra flow, the desired incremental increase is nearly instantaneous upon injection. This type of application is utilized in many oil-producing regions. The cost is minimal per incremental barrel produced.
3. Pressure reduction
   - Reduced operating pressure to handle corrosion problems
   - Reduced operation pressure due to maintenance
   - De-bottlenecked connected platform infrastructure

Pipeline operations depend on pumping pressure as their lifeline to deliver flow capacity. Due to aging systems, corrosion, abrasions, or a pipeline bottleneck, pressure constraints become a pipeline's Achilles' heel. DRAs allow continuous operation within the constraints of maximum allowable operating pressures without sacrificing capacity.

Drag-reducing additives are used to manage pressure constraints and/or enable typical capacity at lower pressures.

4. Energy savings

Drag-reducing additives are commonly used to maintain a pipeline's flow rate while bypassing a pumping station that is down temporarily for service or repair. However, drag-reducing additives can be used as a permanent pump replacement.

This inventive approach has allowed the operator to bypass half of the pumping stations on the system and significantly lower overall energy use. This "energy hedge" delivers a significant seven-figure dollar savings each year, reduces maintenance and repairs costs, and provides a "green" alternative to energy usage (Table 4.1).

### 4.2.6.1.6    Effects of Drag Reducer on End-Use Equipment

Of major importance is the effect of drag reducers on the equipment and operations that ultimately the end users must be responsible for. Previous work on drag reducers has been accomplished by Conoco on its own Conoco Drag Reducer (CDR). Both Conoco and Trans-Alaska pipeline system (TAPS) ran a set of bench tests prior to use of CDR in TAPS. In ASTM foaming tests, addition of CDR had no significant effect upon the results. In desalting test, no adverse results were seen on high-temperature emulsion stability. High-temperature heater tests were run and no heat exchanger fouling tendencies were observed. Coking of crude residue was evaluated with concentrations of CDR of up to 2000 ppm. It is probable that the tank-filling procedures, in particular the filter separator, will fully degrade the pipeline drag reducer additive by the time it is added to the vehicle tanks (Laura and Tim, 2010).

The results of the evaluation of certain parameters such as the number of revolutions, thrust, exhaust temperature, specific fuel consumption, and endoscopic tests showed no significant influence of the Pipeline Drag Reducer on the operation of the engine.

The following conclusions are made on the effects of drag reducers on end-use equipment:

- Use of drag-reducing additives will result in more carbon deposition in fuels.
- The use of pipeline drag-reducing additive, in particular CDR 102M, in turbine fuels will result in increased carbon deposition in engine fuel injection nozzles and manifolds where fuel-wetted wall temperatures are 450°F or greater.
- The increased carbon deposition associated with the use of some drag-reducing additives will result in increased maintenance requirements for engines.
- Future engine trends indicate that increased bulk fuel and engine nozzle wetted wall temperatures will make the use of drag-reducing additives such as CDR 102M even more unattractive from a thermal stability standpoint.

**TABLE 4.1**
**Application of Drag Reducers/ Flow Improvers Polymer Additives**

| Problem Solved | Location | Liquid | Pipe Diameter (Inches) | Miles | Throughput MBPD | % D.R. Achieved | Approx. Flow Increase (%) |
|---|---|---|---|---|---|---|---|
| Increase oil production | Alaska | Crude | 16 | 28 | 120 | 38 | 30 |
| Avoid pump station construction | Alaska | Crude | 48 | 150 | 1450 | 22 | 15 |
| Increase oil production | Offshore Gulf Coast | Crude | 6 | 2 | 6 | 16 | 10 |
| Increase oil production | Far East | Crude | 12 | 14 | 24 | 28 | 20 |
| Increase oil production | Middle East | Crude | 40 | 600 | 720 | 22 | 15 |
| Reduce transport cost | Southwest US | Crude | 16 | 425 | 70 | 19 | 12 |
| Increase oil production | Far East | Crude | 14 | 96 | 60 | 26 | 18 |
| Increase oil production | Offshore Gulf Coast | Crude | 10/12 | 80 | 44 | 38 | 30 |
| Transport bottleneck | Gulf Coast | Crude | 6 | 11 | 134 | 28 | 20 |
| Reduce transport cost | Mid-West US | Crude | 8 | 31 | 16 | 16 | 10 |
| Seasonal demand | Mid-West US | Diesel | 6 | 51 | 36 | 38 | 30 |
| Seasonal demand | Northeast | Gasoline | 8 | 50 | 30 | 28 | 10 |
| Product mix | Southwest US | Gasoline | 8 | 50 | 30 | 40 | 30 |
| Winter to summer rate | Mid-West US | No.2 fuel oil | 8 | 60 | 25 | 38 | 28 |
| Anticipated expanded market | Gulf Coast | Diesel | 12 | 50 | 95/105 | 35 | 20 |
| Proration | Eastern US | Crude | 6 | 40 | 15 | 48 | – |
| Mechanical reliability | North Sea | Crude | 16 | 12 | 145 | 38 | 15 |
| Production surge | Gulf Coast | Crude | 12 | 28 | 85 | 39 | – |
| Planning evaluation | Mid-West US | Crude | 8 | 28 | 35 | 35 | 15 |
| Increase allocation | Southwest Asia | Crude | 24/36 | 82 | 450 | 21 | 8 |
| Alternative transport | North Sea | Crude | 14/16 | 30 | 160 | 26 | 18 |
| System bottleneck | Gulf Coast | Crude | 28 | 117 | 495 | 23 | 7 |

*Source:* Dean, Hale. June 1984. "Special Report–Slick Way to Increase Capacity." *Pipeline & gas journal* 17–19. (*Source:* https://smartpigs.net/pigging-products.html)

### 4.2.6.2 Wax Crystal Modifier Additives

Wax problems occur in lubricating oils, distillate fuels, residual fuels, and crudes. Use of additives in lubricating oils began in the 1930s.

When a waxy fluid is cooled below its cloud point, wax crystals form and begin to agglomerate. As the temperature goes down, crystal accumulation reaches a point when a loose gel structure is formed. The gel can be broken down by shear action but has a tendency to reform on standing. Adding of wax crystal modifiers can transform the growth and size, and reduce the tendency of the crystals to bond to each other. All this reduces the temperature at which the gel structure is built up during the cooling of an additive-treated waxy fluid, making it easier to push through a pipeline. Such materials are known as "pour point depressants."

Considerable and expensive handling problems can result if waxy crudes are not treated. These include:

- The wax deposits on the internal surface of a pipeline could cause serious localized corrosion (such as pitting corrosion and crevice corrosion), which is called as under deposit corrosion.
- Loss of pipeline capacity as wax precipitation builds up on internal pipe walls.
- Development of no-flow conditions in pipelines.
- Inability to restart flow after crude gels.

Waxy crystal modifier additives are practical, cost-effective solutions. They require only relatively modest capital and operating cost to apply. The potential benefits include:

- Unproblematic handling or use of waxy crudes in pipelines
- Assured restartability
- Minimal low-temperature pumping energy

#### 4.2.6.2.1 *Replacement for Wax Crystal Modifier Additives*

There are some alternatives to wax crystal modifier additives, but all are considered significantly higher in cost than using them as flow improvers. The alternatives for wax crystal modifiers include:

- Dilution of waxy crude with less waxy oil or other solvents.
- Heating the waxy oil to temperatures above wax crystallization. These require heaters or furnaces and insulated or heat-traced pipelines, both of which are very costly.
- Installation of additional insulation.

### 4.2.6.3 Heavy and Asphaltic Crudes

Asphaltic and heavy crudes do not react to wax crystal modifier additives or drag-reducing additives. Pipeline emulsions can eliminate the common and expensive practice of cutting heavy, viscous crudes with light hydrocarbon diluents. Saved are the costs of diluent purchase and transportation as well as some of the costs associated with pipeline insulation and/or heating, which often must be used, at significant expense, for transporting heavy oil.

Water-continuous emulsions of even extremely heavy oil have viscosities close to that of water, thus achieving more efficient viscosity reduction than hydrocarbon diluents. This reduces the energy required to transport the fluid and eliminates the need for insulated or heated pipelines. These emulsions can be used for transporting heavy crude over long distances or through infield gathering systems.

Long-distance pipeline emulsions are designed to remain very stable over extended periods of time and at different shear conditions. In addition, while stable during transportation, pipeline emulsion can be broken by the addition of demulsifiers and/or heat at the destination point. Emulsions used in infield gathering systems are deliberately designed to be easily broken at the field treating plant. They can provide an attractive alternative to trucking heavy crude from the wellhead to the treating plant.

### 4.2.7 EMULSION BREAKERS

An emulsion is a mixture of two or more normally immiscible liquids like oil and water. In a typical water-in-oil emulsion, tiny bubbles of water are interspersed within the oil and are stabilized by surface forces. An emulsion breaker destabilizes the bubbles, releasing the liquid which is then forced to form a large bubble, using electrostatics and settle out by gravity.

Many of the chemical additives discussed here apply only to oil processing, with just a few applicable to gas processing, such as corrosion inhibitors and hydrate inhibitors.

### 4.3   DEHYDRATION

Dehydration is the most commonly applied measure to protect against internal corrosion in gas pipelines (and in liquid pipelines that contain oil with free water or other electrolytes). Dehydration removes condensation and free water that, if permitted to remain, would allow internal corrosion to occur at points where water droplets precipitate from the gas stream to either form liquid puddles at the bottom of the pipe or adhere to the top of the pipe. Where the gas stream is usually dry, topside corrosion rarely takes place. Complete dehydration is very effective, but because the systems are neither 100 percent effective nor 100 percent dependable, there always is the potential to introduce water and other electrolytes into a gas pipeline.

Gas dehydration is used to remove water from the gas. The water content of a gas stream is expressed as:

- Weight of water per volume of gas (mg/Sm³)
- Dew point at a reference pressure (temperature at which the water vapor condenses within the gas)
- A concentration in parts per million of water in gas (ppm)

The following determines the water-holding capacity of the gas:

- Higher temperature.
- The higher the temperature, the higher the gas's ability to hold more water.
- Lower pressure.
- Low pressure increases the water-holding capacity of the gas.
- Presence of $CO_2$ and $H_2S$ in the gas at high pressures.
- A gas with these impurities will hold more water at higher pressures, and correction should be made for this when calculating the water content of such a gas stream, especially when the gas mixture contains more than 5 percent $H_2S$ and/or $CO_2$ at pressures above 4800 kPa.

Typical dehydration strategies for both the upstream and downstream section of the gas value chain are:

- Glycol dehydration at gas processing plant (see Figure 4.8)
- Partial dehydration at the wellhead and later additional steps to meet contract specifications
- Chemical injection at the wellhead with later dehydration at the central delivery point
- Full and complete dehydration at each wellhead

## 4.3.1 Reason for Dehydrating the Gas

Dehydration is used to control the water dew point. This is important to meet the product water dew point specification and to avoid operational problems.

Water condensation can cause:

- Corrosion: $CO_2$ and $H_2S$ corrosion in carbon steel pipelines
- Hydrates

## 4.3.2 Common Gas Dehydration Methods

Main technologies used for gas dehydration are:

1. Glycol dehydration (physical absorption) (MEG, DEG, TEG, TREG)
2. Adsorption on solid bed (e.g., molecular sieves)
3. LTS (low-temperature separators)with glycol injection system (TEG)

### 4.3.2.1 Glycol Dehydration

The process is carried out by the absorption of the water vapors in a glycol solution in a contactor tower. Highly concentrated glycol solutions (TEG) are used to physically absorb the water from the gas, and the glycol solution is then regenerated so it can be used again.

A typical dehydration unit is presented in Figure 4.8. The design can vary from plant to plant.

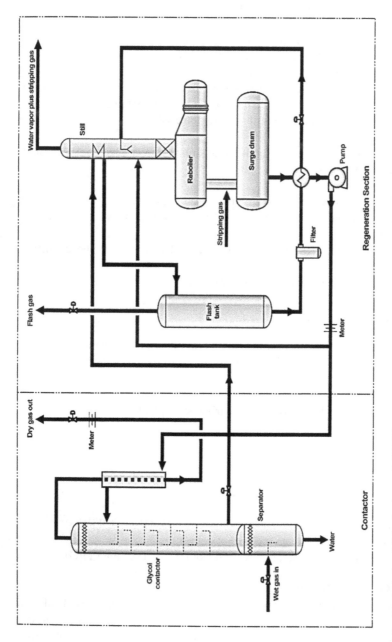

**FIGURE 4.8**　Typical glycol dehydration unit (GDU).

### 4.3.2.2   Adsorption on Solid Bed (e.g., Molecular Sieves)

Separation processes by adsorption uses a solid bed material with a large specific surface area. There are several solid desiccants that possess the physical characteristic to adsorb water from natural gas (see Figure 4.9).

The solid desiccants for commercial use are divided into three categories:

- Gels (alumina or silica): Silica gel is generally used for dehydration and i-C5+ recovery from the natural gas stream. When silica gel is used for dehydration, the achieved water dew point is approximately −60°C.
- Alumina (activated aluminum oxide): Alumina as solid desiccant reaches water dew point down to −73°C.
- Molecular sieves (alumina-silicates): Molecular sieves have the highest water capacity and can get down to a water dew point of −100°C.

The adsorption on solid beds process requires two (or more) vessels, with one of the vessels removing water while the other is being regenerated and cooled.

Usually one adsorption cycle can last from 8 to 24 hours.

In general, four methods are commercially available in order to carry out the regeneration of saturated desiccant material:

- Temperature-swing adsorption (TSA)
- Pressure-swing adsorption (PSA)
- Inert-purge stripping

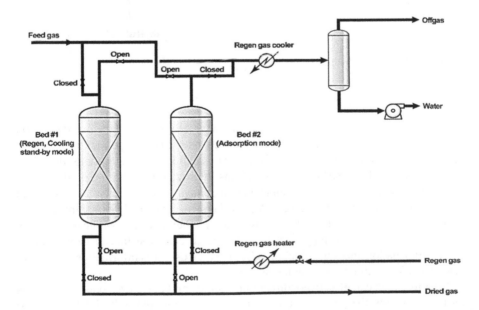

**FIGURE 4.9**   Typical gas dehydration, by adsorption process.

### 4.3.2.3   Low Temperature Separator (LTS) with Glycol Injection System

A low-temperature separator with glycol injection system is used if the water or HC dew point cannot be met and also to inhibit hydrate formation. It is typically used when it is not cost-effective to install a full GDU (gas dehydration unit) or a solid adsorption unit.

The gas is cooled to condense both water and hydrocarbons (NGL). Cooling is performed using mechanical refrigeration, J-T effect, or Turbo-expander. The water and hydrocarbon dew point are determined by the operating temperature of the cold separator. The glycol injection prevents the formation of hydrates which can block the equipment. This process is illustrated in Figure 4.10.

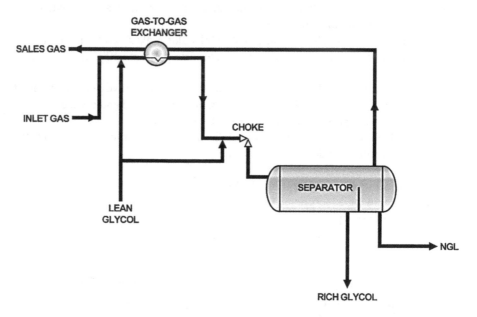

**FIGURE 4.10**   Gas dehydration module.

## 4.4   CLEANING PIGS

"Pig" is an acronym for "pipeline inspection gauge." Pigging is an in-line inspection (ILI) method in which devices known as pigs are inserted into pipelines to clean and/or inspect the pipeline. A pig can perform various maintenance operations, including cleaning the pipeline and making inspections to provide information on the condition of the pipeline and the extent and location of any problem, such as corrosion and isolation of a pipeline, by plugging it. A pig is launched from a pig launching unit and is driven by the product fluid. It is received at the other end by a pig receiver device.

The main benefits of pigging a pipeline are to clear the line of debris and to gather information about the pipeline. The user then uses this information to map out defects to aid repair crews in maintenance.

Cleaning pigs can effectively direct both liquids and corrosive solids to pig traps for removal from the pipeline. Routine maintenance pigging will direct any liquid pools away from low points and, if performed properly, will direct any liquid pools out of the entire pipeline. Cleaning pigs also will move the solids and remove them from the pipeline through the pig trap at the end of the pipeline (Figure 4.11).

**FIGURE 4.11**    Cleaning pig.

These pigs are specially made to remove sedimentation and buildup that can impede the flow of materials.

There are two ways of cleaning a pipeline:

1. Mechanical cleaning

    The generally approved and accepted practice among pipeline operators for removing deposits in a pipeline is by *mechanical pigging*. The pig is repetitively sent through the pipeline to swap deposits from that pipeline until hardly any deposit can be found in the pig receiving station. It is, however, hard to determine if the latter condition implies that the pipeline is clean.

2. Advanced chemical cleaning

    "Advanced chemical cleaning" is rapidly becoming an industry standard. Chemical cleaning in conjunction with the use of mechanical pigs removes a greater volume of debris with fewer runs. Chemical cleaning, by definition, means the use of liquid cleaners mixed in a diluent (water, diesel, methanol, isopropyl alcohol, and the like) to form a cleaning solution that can be pushed through a pipeline by pigs. For example, the removal of hydrocarbon hydrates from a pipeline system involves running a batch of either methanol or glycol through the pipeline, driven by a pig. The size of the batch should be sufficient to depress the hydrate formation temperatures of the maximum anticipated hydrate deposition to below the pipeline temperature as well as allowing for liquid that will be left on the pipe wall and not be pushed forward by the pig. Removal of the hydrates from raw natural gas pipelines, which may not have pigging facilities, is usually achieved by increasing the dosage rate of hydrate point depressant. For a short period, shock dosing, at a rate of approximately five times the normal injection rate, should be undertaken.

    Selection of designed pipeline cleaners should be based upon the following features (Werff, 2006):
    - A neutral pH
    - Deposit permeating and penetrating capabilities

- Original design parameters of the cleaner and its case histories
- Health Safety and Environment (HSE) awareness

## 4.5  BUFFERING

A buffering agent is a weak acid or base used to maintain the acidity (pH) of a solution near a chosen value after the addition of another acid or base. That is, the function of a buffering agent is to prevent a rapid change in pH when acids or bases are added to the solution.

Buffering agents change the chemical composition of fluids that remain in the pipeline and hence could be utilized to prevent internal corrosion.

A buffering agent, such as a mild or dilute alkaline mixture, can significantly reduce the corrosivity of any standing liquid, predominantly by raising its pH value above seven (neutral), so that it turns from acidic to alkaline. Alkaline liquids cause virtually no harm to steel. In general, buffering is not very effective because it is difficult to cover the entire pipe surface.

The use of buffers or other pH-altering chemicals also can modify the environment and possibly eliminate the growth of the bacteria that cause microbial corrosion (Baker, 2008).

# 5 Atmospheric Corrosion

Atmospheric corrosion refers to the corrosive action that occurs on the surface of a metal in an atmospheric environment. It occurs when the surface is wet by moisture formed due to rain, fog, condensation, dew, precipitation, and relative humidity. Atmospheric corrosion occurs at pipe bodies, pipe supports, pipe fittings, aboveground pipe transitions, valve and valve bodies, and so on.

To prevent all forms of atmospheric corrosion, there is a need to specify and consider the conditions of exposure at the design stage. It is far easier and cheaper to select materials and protective systems in advance than to clean up corroded components and apply protection after corrosion has set in. If this does become necessary, thorough cleaning is essential, and this involves removing the corroded material down to the base of the pits. It is difficult and expensive to recondition corroded components, and it is better to take precautions in advance.

Atmospheric corrosion evaluation includes a visual inspection of all aboveground metallic facilities. Signs of atmospheric corrosion include:

1. Discolored and/or peeling paint
2. Evidence of rust or oxidation of the metal surface
3. Any other condition that may require remedial action

For portions of piping and structures exposed to atmospheric corrosion, the following remedial actions apply.

The portion of metallic pipe, fittings, valves, supports, and/or appurtenances exposed to the atmosphere shall be thoroughly cleaned (removing the corroded material down to the base of the pits) and painted or coated with an ultraviolet-light-resistant paint or coating that has been approved for above-the-ground use, such as:

1. Protective coatings
   - Urethane coatings.
   - Two-part liquid-epoxy coating.
   - Wrappings.
   - Temporary protective materials: Temporary protective materials such as grease or slushes, pastes; wrapping papers, which may be greased or waxed, may be applied to the surface to exclude moisture and dust.
   - Paint films: Includes a priming layer containing a corrosion inhibitor, or primers containing metallic zinc particles may be used.
   - Surface coatings of metals.
   - Polymers or vitreous enamels: For example, thick layers of borosilicate glass can give excellent protection against corrosion and oxidation up to about 900°C.
   - Conversion coatings.

**FIGURE 5.1**   Atmospheric corrosion on a pipe.

2. Sacrificial coatings: sacrificial coatings such as zinc, cadmium, or aluminum can be used as a foundation for a paint system (Figure 5.1).

## 5.1   ATMOSPHERIC CORROSION INSPECTION

1. Examine exposed piping for atmospheric corrosion.
2. Examine transition coating or tape for bonding and integrity.
3. Examine support structures for atmospheric corrosion and mechanical integrity.

## 5.2   CAUSES OF ATMOSPHERIC CORROSION

- Moisture (e.g., rain, dew, condensation, high relative humidity [RH]).
- High RH (above 70–80 percent): Corrosion can be speeded up by high relative humidity. Corrosion is slowed down significantly when relative humidity is below 50 percent. Relative humidity is useful only when measured at the surface.
- Salt mist.
- Surface contaminants (dust, sweat residues, soldering fluxes, etc.).
- Atmospheric contaminants (SO, HCl, organic acids).
- High temperature.
- Oxygen.

## 5.3   METHODS OF PREVENTING ATMOSPHERIC CORROSION

To prevent atmospheric corrosion, the following measures should be applied:

- Cleaning and protection from dust
- Drying the atmosphere and preventing condensation
- Purifying the atmosphere
- Use of corrosion inhibitors
- Protective coatings (wrappings, temporary protective materials, paint films, surface coatings of metals, polymers or vitreous enamels, conversion coatings)
- Sacrificial coatings

## 5.3.1 COATINGS

If oxygen, water, and strong electrolytes are required to produce rapid corrosion, one or more of these agents may be excluded by a variety of coatings, ranging from noble metal coating and vitreous enamels through thick polymeric coatings (e.g., polyethylene, nylon, and PVC) and paint films to paper wrappings and thin coatings of polymer or greases (Lloyd, 2003).

A noble metal coating is more corrosion-resistant than the substrate, and it provides protection when it is a pore-free barrier coating. Noble coatings on steel are therefore expected to act strictly as barriers to prevent corrosion of the steel substrate.

## 5.3.2 METAL FILMS

In principle, electrodeposits of metals might provide impervious barriers: Gold, silver, nickel, and chromium deposits are used both for decorative and for protective purposes (Lloyd, 2003). In practice, however, electrodeposits are invariably porous, and are liable to mechanical damage. Exposure of the base metal leads to the damaging combination of small anode and large cathode, so that intense corrosion forms a bulky corrosion product in the pore, which may further damage or detach the coating. Thick deposits produced by high-temperature process ("chromizing," "calorizing," etc.,) are comparatively robust.

## 5.3.3 POLYMER COATINGS

A number of commercial processes are available for applying polymeric materials from either sheet or powder, and comparatively thick, robust coatings can be obtained which give protection against corrosion and mechanical damage. Like metallic barrier layers, they may, however, allow rapid corrosion if portions of the metal are exposed (Lloyd, 2003).

Polymer coating systems can be applied to metal surfaces, ceramics, and synthetic materials to provide anticorrosion protection. However, polymer coatings develop microcracks easily in structural applications, reducing lifespan, so early sensing, diagnosis, and repair of microcracks are important. Incorporating microcapsules into the coating matrix enables the release of a repairing agent rapidly after triggering of crack propagation in coatings, leading effectively to self-healing.

Examples of polymeric coatings include:

- Natural and synthetic rubber.
- Urethane.
- Polyvinyl chloride.
- Acrylic, epoxy, silicone.
- Phenolic resins.
- Nitrocellulose.
- Acrylics and alkyds: Usually used for farm equipment and industrial products requiring good corrosion protection at a moderate cost.
- Polyurethane: applied on conveyor equipment, aircraft, radomes, tugboats, road-building machinery, and motorcycle parts. Abrasion-resistant coatings

of urethanes are applied on railroad hopper cars, and linings are used in sandblasting cabinets and slurry pipes.
- Nylon 11: Provides attractive appearance as well as protection from chemicals, abrasion, and impact.

They are temperature-resistant up to approximately 535°F (280°C). Applying a polymeric coating to a metallic surface increases the ionic resistance.

### 5.3.4 Vitreous Enamels

Vitreous enamels are glass-like coatings that can be fused onto the surface of metals, ceramics, and glass. Comparatively thick layers of borosilicate glass can give excellent protection against corrosion and oxidation up to about 900°C. Again, mechanical damage leads to failure of protection against corrosion (Lloyd, 2003).

### 5.3.5 Conversion Coatings

Conversion coatings are coatings for metals where the part's surface is subjected to a chemical or electrochemical process by the coating material that converts it into a decorative or protective substance. Conversion coatings are normally formed by immersing or spraying an aqueous solution that activates a metal surface, dissolves any existing surface oxide, and replaces it with a thin mixed-metal oxide coating.

Chemical reactions to produce layers of corrosion-resisting scales (particularly phosphate and chromate) can produce a wide range of coatings, suitable both for enhancing the corrosion resistance of the metal or as a preparation for painting. Anodizing treatments for aluminum may incorporate phosphate or chromate ions, and so-called "chemical anodizing" may be used to enhance the resistance of aluminum to atmospheric attack. Chromate treatments are also used to improve the resistance of zinc and cadmium plating (Lloyd, 2003).

Overall, there are three categories of conversion coating:

1. Oxide conversion coating: This type of coating is an anticorrosion product that is ultrathin and offers good adhesion. In such cases, oxide treatments can be performed through electrochemical, heat, or chemical reactions. The best examples of oxide coatings include chemical baths, black oxide, and anodizing.
2. Phosphate conversion coating: This is produced by the chemical conversion that exists on a metal substrate in order to produce a highly adhesive phosphate coating. The crystals that may build up on the surface include manganese, zinc, and iron. Out of these three, phosphate of manganese is the best type of coating for wear applications. It is ideal for low alloy metals, cast iron, and carbon steel. It is one of the most beneficial forms of coating material that is nonmetallic.
3. Chromate conversion coating: This is comparable to phosphate coating since it is created by chemical conversion. It is formed through the interaction of chromium salts or chromium acid with water solutions. This type of

coating is applicable to zinc, aluminum, magnesium, and cadmium. This coating normally provides superior corrosive resistance and is broadly used in giving protection to usual household products such as hinges, screws, and other hardware items.

## 5.3.6  PAINTING

Paints provide easily applied and versatile organic coatings which can be adapted to a wide range of requirements. They do not act by excluding water and oxygen; their action depends partly on excluding strong electrolytes and partly on the fact that paint materials usually contain corrosion inhibitors, often in the form of heavy-metal organic salts.

Paint films are porous, and they are liable to mechanical damage. Even if corrosive salts penetrate breaches in the paint, however, the inhibitive species will often suffice to previous serious attack on the underlying metal. If the damaged area is too large, or the contamination too heavy, and corrosion proceeds, the cathodic alkali may soften and detach the paint film, leading to the formation of a blister that remains wet long after the rest of the surface has dried.

A successful protective system should therefore include a priming layer containing a corrosion inhibitor and preferably a layer of metal that can protect the underlying surface by sacrificial action. Steel to be painted may therefore be aluminum- or zinc-sprayed, or primers containing metallic zinc particles may be employed.

Thorough cleaning of the metal surface is extremely important, especially if it has previously been corroded. Abrasive treatment should be sufficiently severe to remove the whole of any previously formed corrosion product. It has been shown that on rusted steel, crystals of ferrous sulfate are present at the base of quite deep pits. Since paint films are permeable to both water and oxygen, corrosion cells can be set up under the paint unless the preparation is sufficiently thorough (Lloyd, 2003).

## 5.3.7  SACRIFICIAL COATING

A sacrificial coating is more active than the substrate, and it provides protection first as a barrier and secondly as a sacrificial coating; that is, the coating cathodically protects the substrate at exposed edges and pits (holes) through the system.

Zinc, cadmium, and aluminum are examples of sacrificial coatings and can protect steel by sacrificial action: being anodic with respect to iron, they hold the potential of the latter at a value at which it will not corrode. Since the corrosion product is usually highly protective, these metals build up a surface layer that limits corrosion and prolongs the life of the coating (Lloyd, 2003). These coatings can be used alone or as a foundation for a paint system and chemical conversion coatings can enhance their effect.

## 5.3.8  TEMPORARY PROTECTIVES

Apart from wrapping papers, which may be greased or waxed, temporary protective materials can be applied to metal surfaces to exclude moisture and dust.

They may take the form of greases or slushes, pastes, thin films applied with solvents, or thick films of soft thermoplastic materials that give mechanical protection. All these types of protective materials may contain corrosion inhibitors (Lloyd, 2003).

### 5.3.9 DESIGN

The normal considerations of design against corrosion apply to atmospheric exposure. Parts should be designed to avoid trapping moisture in hollow sections, crevices, or joints, all of which may lead to locally increased time of wetness. Suitable drain holes should be provided for hollow sections. Structures should also be designed to avoid setting up "poultices" of moisture-retaining debris and to prevent insulating materials from becoming soaked with water. Many corrosion-inhibiting putties, slurries, pastes, and impregnated tapes are available for avoiding vulnerable design features (Lloyd, 2003).

Fastenings should be of the same material as the components they secure, or if this is impracticable, they should not be more anodic; "small cathode/large anode" is better than the reverse. It is usually better to avoid even small differences in material, if possible.

### 5.3.10 CONTROL RELATIVE HUMIDITY

Most atmospheric corrosion can be prevented by maintaining relative humidity (RH) below 60 percent. Desiccators and dehumidified stores can therefore be used for storage. In storerooms and warehouses, it is important to maintain the air temperature at a reasonable level and to avoid large variations in temperature; a fall in temperature overnight or at the weekend may lead to heavy condensation of moisture. Condensation may also occur if massive metal parts are placed while cold into a warm room if the air is not saturated at the prevailing temperature. Stoves and gas heaters must be provided with adequate flues. The air in store cupboards may be dried by the use of desiccants or by a refrigerating plant. For special purposes, including display cabinets, where it is essential to ensure immediate access to complex equipment, refrigerated surfaces may be the most practical means of protection (Lloyd, 2003).

### 5.3.11 PACKAGING

Packages may employ desiccants: the two most common agents are silica gel and activated alumina, both of which are noncorrosive and can be regenerated by heating.

Desiccated packages need careful design to take account of the permeability of the wrapping material to water vapor, the water content of the interior of the package, and the absorptive capacity of the desiccant.

The metal parts should be carefully cleaned before packaging to minimize damage if the package is breached or the desiccant becomes exhausted after prolonged storage. It should be remembered that some packaging materials may contain appreciable quantities of soluble chloride or sulfate, or may liberate corrosive vapors. The choice and specification of materials should consider these factors. Maximum permissible levels are:

- Chloride (as NaCl) 0.05 percent
- Sulfate (as sodium sulfate) 0.25 percent
- pH of water extract 5.5 to 8.0

Desiccated packages are often the best choice for delicate scientific equipment since this can be packed in assembled form without greasing or dismantling, and therefore without the need for cleaning and reassembly (Lloyd, 2003).

### 5.3.12 ATMOSPHERIC CONTROL

Unforeseen problems sometimes arise because variations in temperature lead to heavy condensation of moisture on steel parts, or because of some source of pollution. Troubles of this kind can often be overcome only by identifying and removing the causative agent.

Degreasing solvents and the equipment used for recovering them frequently cause outbreaks of rusting in engineering plants, since heating or chlorinated hydrocarbons may produce hydrochloric acid either in the air or in the solvent. Water extracts should be checked for acidity (Lloyd, 2003).

Fingerprints and traces of soldering or welding fluxes occasionally cause rusting in moist air, and precautions may be needed to prevent or remove contamination, for example, by the use of noncorrosive fluxes.

PVC insulating materials and other polymers may produce corrosive vapors that lead to heavy attack on metal parts. Some types of wood used in packaging produce acid vapors.

Ozone, produced by sparks from electrical contacts, may attack polymers to produce organic acids.

Sulfur compounds liberated by some synthetic rubbers may cause heavy corrosion of silver relay contacts.

Organic acids used in dyeing components made of plastics may be liberated during the life of the equipment and cause severe corrosion.

Traces of ammonia in the air may cause stress corrosion cracking of brass or copper.

## 5.4   ATMOSPHERIC CORROSION REPAIR

### 5.4.1 SURFACE PREPARATION

The damaged surface(s) shall be thoroughly clean, dry, and free from condensation, moisture, dust, oil grease, rust, dirt, and other contaminants before the application of repair material.

Surface preparation, whenever possible, shall be carried out by dry blast cleaning. Where dry blast cleaning is not feasible due to limited access, risk of damage to equipment, light gauge steel, proximity to electrical or instrumentation components;

hand or power tool cleaning shall be applied. This shall be followed by solvent cleaning prior to painting.

The use of brushes or rollers for touch-up/repair on localized damaged surfaces where proper coating by spray application is not feasible may be used upon client approval.

When using brushes, ensure that a smooth coat, as uniform in thickness as possible, is obtained with no deep or detrimental brush marks.

Anti-rust primer/zinc coating shall be used to coat any rusty area hard to reach by powder brush, but make sure loose rust is removed by sandpaper before primer application. After that, use clean rag or cloth to remove the foreign matter from the surface before applying the paint/coating.

Where the damaged surface being repaired lies adjacent to a previously coated surface, the cleaning shall extend to the surrounding coating by a minimum of 25 mm on all sides and the edges shall be "chamfered" to ensure continuity of the patch coating.

## 5.4.2   RECOATING

All existing/parent coating shall be brush-blasted or thoroughly abraded before recoating with a material of the same formulation as the existing/parent coat.

Recoating over existing coatings of a different type or formulation shall not be permitted unless approved by both purchaser and manufacturer. Ensure compatibility with the existing/parent coating.

## 5.4.3   INSPECTION

Inspection must be carried out to ensure that the requirements of the specification are being met. All coating repairs shall be inspected for electrical continuity and shall be free of holidays. Holiday testing should be performed to test for discontinuities in the coating.

Coating repairs not meeting specified criteria shall be stripped and re-coated, or if possible, repaired according to an accepted procedure.

## 5.4.4   HEALTH AND SAFETY

The following health and safety measures must be observed:

1. Because of toxicity, there should be restrictions on the use of certain materials and components.
2. Noise produced by processes such as blast cleaning must be kept below a level that would damage hearing.
3. Explosion hazards must be avoided by providing ventilation to remove flammable solvents and/or dust.
4. Operators must be protected against such hazards by the provision of protective clothing, fresh air masks, ear muffs, and so on.
5. Provision of adequate ventilation/extraction must be present.

### 5.4.4.1   Environmental Protection

Use paints and coatings that are "environmentally friendly" and thereby minimize damage to the atmosphere. Coating/painting shall be free from volatile organic compounds (e.g., solvents).

Waste disposal must comply with client security and environmental policy.

# 6 Stray Current Corrosion

Stray currents are constituents flowing in the electrolyte from external sources. Any metallic structure buried in soil, such as a pipeline, represents a low-resistance current path and is therefore fundamentally vulnerable to the effects of stray currents.

## 6.1 STRAY CURRENT SOURCES

- HV DC and AC Power Transmission system
- Electrical train
- Impressed current cathodic protection (ICCP) system DC

## 6.2 STRAY CURRENT CORROSION PREVENTION

Stray current corrosion prevention is described in Sections 6.2.1–6.2.3.

### 6.2.1 CONSTRUCTION TECHNIQUE

The pipeline shall not follow an overhead power lines in parallel.

### 6.2.2 CORROSION AND PREVENTION OF DC STRAY CURRENT

1. The current out of design or regulated circuit is called stray current. If stray current flows into buried metal and then from metal into earth or water, intensive corrosion will occur at the place where the current flows out. It is usually called galvanic corrosion. Its features are as follows:
   - Intensive corrosion
   - Corrosion is concentrated on the local
   - Corrosion is often concentrated on the defects of coating, if there exists corrosion-resistant coating

Metallic pipelines that are interfered with by stray current could have pitting corrosion in a short time (Figures 6.1 and 6.2).

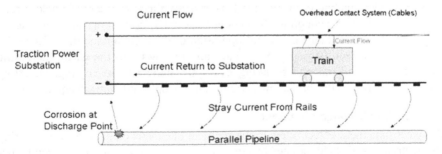

**FIGURE 6.1** DC interference. (Source: Tong, Shan. 2015. *Cathodic protection*. Training document, Ghana: Sinopec.)

**FIGURE 6.2** Stray currents—AC interference. (Source: Tong, Shan. 2015. *Cathodic protection*. Training document, Ghana: Sinopec.)

2. The most practical and effective way to prevent stray current interference is the electric drainage method. This method artificially leads interference current back to the interference source or flows it back to a rectifier through a regression line. This method requires connection between the pipeline and the regression line. This way of preventing pipeline galvanic corrosion is called the electric drainage method. Common drainage methods include the direct method, the polarity method, the compulsive method, and the grounding method, as shown in Table 6.1.

## 6.2.3 AC INTERFERENCE HAZARD AND PROTECTION

### 6.2.3.1 Electric Field Effect

Pipeline-earth potential is slightly increased by the electrostatic field of the high-voltage lines and metallic pipelines, through coupling of distributed capacitance (see Figure 6.3). However, this effect is so small that it could be ignored.

### 6.2.3.2 Earth Electric Effect

If the pipeline is buried in soil whose earth potential gradient changes greatly, the increase of the pipeline-earth potential caused is called earth electric field effect, mainly referring to the coupling phenomenon caused by current in soil. A grounding loop could be ignored during normal operation of the high-voltage lines. However, when there is a fault, strong short-circuit current flows into the earth, increasing pipeline-earth potential and breaking down the anticorrosive coating. Diverting potential generated after breakdown could injure people and damage equipment.

### 6.2.3.3 Electromagnetic Effect

The electromagnetic effect refers to the physical phenomenon caused by alternating the electromagnetic field radiated by current-carrying, conductor-cutting metal pipeline. Induced voltage and current in pipeline is the function of current, frequency, operating methods, and other factors of the high-voltage lines.

**TABLE 6.1**
**Drainage Protection Methods**

| Methods | Direct Drainage | Polar Drainage | Compulsive Drainage | Grounding Drainage |
|---|---|---|---|---|
| Schematic diagram | | | | |
| Application conditions | 1. There is a stationary anodic section on the interfered pipeline. 2. Around grounding electrode of the DC power supply station or negative regression line. | Pipe-to-soil potential on the interfered pipeline has a positive and negative alternation. | Small potential difference between pipeline and track. | Cannot directly conduct electric drainage to interference source. |
| Advantages | 1. Simple and economic. 2. Favorable effect. | 1. Simple installation. 2. Wide application range. 3. Power-free. | 1. Wide protection range. 2. Applicable to special occasions where other electric drainage methods can't be applied. 3. Provides a cathodic protection when train car is out of service. | 1. Wide application range and is applicable to any occasions. 2. Little interference to other facilities. 3. It can provide part of the cathodic protection current (when using sacrificial anode grounding). |

*(Continued)*

## TABLE 6.1 (CONTINUED)
## Drainage Protection Methods

| Methods | Direct Drainage | Polar Drainage | Compulsive Drainage | Grounding Drainage |
|---|---|---|---|---|
| Disadvantages | Limited application range. | Protection performance is poor when the pipeline is far away from the track. | 1. Accelerate galvanic corrosion of the track. <br> 2. There is great influence on potential distribution of the track. <br> 3. Power supply is necessary. | 1. Less efficient. <br> 2. Auxiliary earth bed is necessary. |

*Source:* Tong, Shan. 2015. *Cathodic protection.* Training document, Ghana: Sinopec.

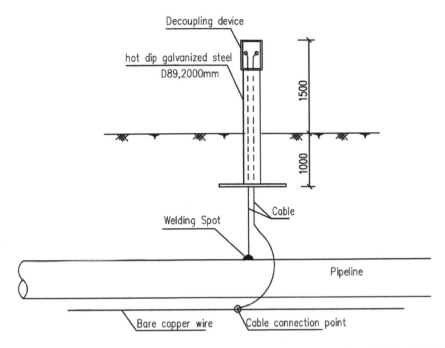

**FIGURE 6.3** Stray currents—interference draining. (Source: Tong, Shan. 2015. *Cathodic protection*. Training document, Ghana: Sinopcc.)

### 6.2.3.4 Protection

The protection measures for buried pipelines are:

- Strengthen the anticorrosive coating quality in places where earth electric field interference exists.
- Use grounding treatment for ground pipelines or pipelines under construction.
- Discharge induced AC on pipelines to the earth.
- Conduct segmented insulation using insulating flange at places where induced AC is not easily discharged.

# 7 Case Study

## 7.1 SITUATION

The East-West (EW) pipelines span 1204 kilometers between "A" and "Y" (AY pipeline) and are served by 11 pump stations distributed according to the hydraulic requirements of the AY pipeline system. The East-West Pipeline crude oil pipeline system consists of two large-diameter parallel pipelines (48-inch AY-1 and 56-inch AY-1 loop) running from "A" plants in the Eastern province to the "Y" crude oil terminal on the seacoast. Currently, the 56-inch AY-1 loop (AY-1L) is the only remaining pipeline transporting four crude grades to "Y" crude oil terminal. About 173 km of the 56-inch AY-1L pipeline is coated with PE and has been in service since 1989.

Currently, the polyethylene (PE) coating of the AY-1 pipeline is damaged (Figures 7.1 and 7.2).

This 173 km of EW pipelines from KM-154 to KM-327 is protected by eight CP systems located approximately every 30 km with total current output of 284 amperes. There are two sections of low potential, as follows:

- KM-164 to KM-184 with minimum CP potential of -720 mV at KM-174. This could be due to high soil resistivity, coating damage, and low CP current output.
- KM-261 to KM-322 with minimum CP potential of -443 mV at KM-310. This could be due to coating damage and low CP current output.

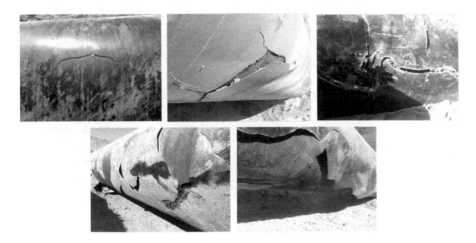

**FIGURE 7.1**  Damaged AY-1 polyethylene (PE) coating.

**FIGURE 7.2**   Damaged pipeline coating.

### 7.1.1   EXTERNAL CORROSION COUPONS

External corrosion coupons (surface area of 44 sq. cm) were installed at a few locations along this pipeline section. Table 7.1 presents a list of the coupon locations along with their measurements (Figure 7.3).

---

**TABLE 7.1**
**External Corrosion Coupons**

| P/L | KM | Coupon ON | Instant OFF | Native | Polarization Decay | Protection Status |
|-----|-----|-----------|-------------|--------|--------------------|-------------------|
|     |     | −mV | −mV | −mV | mV | Protected/Not protected |
| AY-1L | 308 | 520 | 491 | 267 | 224 | Protected |
| AY-1L | 314 | 788 | 683 | 407 | 276 | Protected |

---

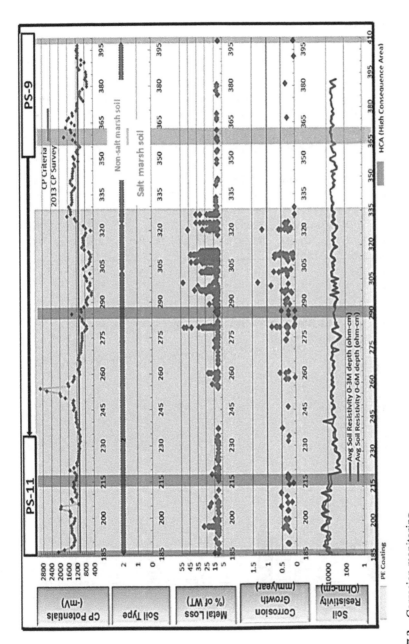

**FIGURE 7.3** Corrosion monitoring.

## 7.1.2 History of Metal Loss

**TABLE 7.2**
**History of Metal Loss**

| No. | Type | Length (km) | Start (KM) | End (KM) | Number of External Indications in the Last ILI Run | Maximum Metal Loss % | Average Metal Loss % |
|-----|------|-------------|------------|----------|-----------------------------------------------------|----------------------|----------------------|
| 1 | Valley | 0.966 | 278.797 | 279.763 | 21 | 79 | 29 |
| 2 | Salt marsh soil | 4.1 | 293.912 | 298.012 | 20 | 67 | 29 |
| 3 | Valley | 0.768 | 302.601 | 303.369 | 9 | 41 | 28 |
| 4 | Small valley/ rocky | 4.1 | 306.301 | 310.412 | 67 | 43 | 29 |
| 5 | Valley | 0.234 | 319.959 | 320.193 | 8 | 34 | 26 |
| 6 | Valley | 0.353 | 322.624 | 322.977 | 8 | 45 | 25 |

AY-1L-6 Number of Metal Loss features KM-185 to KM-327

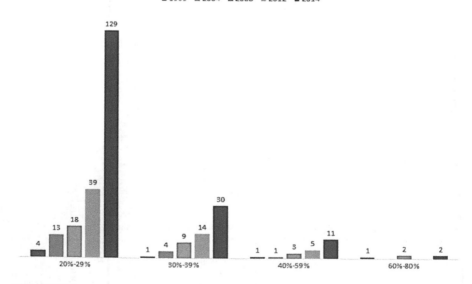

**FIGURE 7.4**    Metal loss.

## 7.2 STEPS INVOLVED

This investigation involved:

1. Visits to the pipeline sites
2. Collecting coating samples
3. Conducting laboratory testing on the coating samples:
   - FTIR Spectroscopy
   - Microscopy
   - Differential Scanning Calorimetry (DSC)
   - X-Ray Diffraction (XRD)
   - Electrochemical Impedance Spectroscopy (EIS)
4. Soil samples testing
5. CP data analysis

### 7.2.1 LABORATORY TESTING: MAJOR FINDINGS

The laboratory results showed that coating degradation and lack of adhesion appear to be the main contributors to the failure of the polyethylene (PE) coating. The results have shown that coating degradation could be due to water vapor and oxygen permeation.

Fourier-transform infrared (FTIR) analysis suggests that the outer surface of the PE sheet has been oxidized by oxygen/air, while the inner surface remained unaffected.

The presence of iron oxide, silicon, and sodium at the interface between the PE sheet and the pipe could be an indication of poor surface preparation during coating application that leads to poor adhesion at these locations.

### 7.2.2 ELECTROCHEMICAL IMPEDANCE SPECTROSCOPY (EIS)

EIS readings were relatively low near the cracks that may suggest a coating degradation at those locations. This did not give a strong confirmation that the coating material has failed due to material degradation. $R_p$ value of higher than $10^8$ means good protection is provided by the coating (Figures 7.5 through 7.8 and Table 7.3).

**FIGURE 7.5** Electrochemical impedance spectroscopy (EIS).

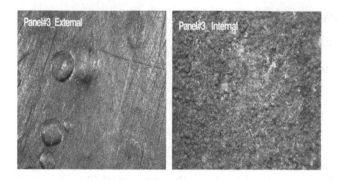

**FIGURE 7.6**  SME images.

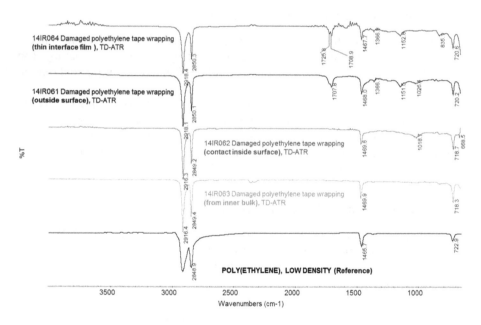

**FIGURE 7.7**  Fourier transform infrared (FTIR).

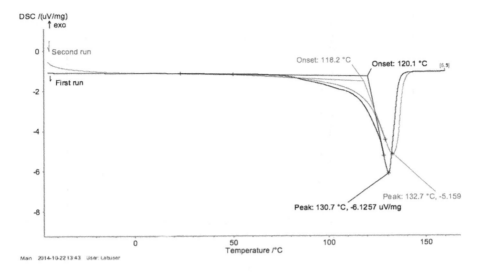

**FIGURE 7.8**  Differential scanning calorimetry (DSC).

---

**TABLE 7.3**
**Electrochemical Impedance Spectroscopy (EIS) Reading**

| Panel # | External Side ($\Omega$) | Pipe Side ($\Omega$) |
|---|---|---|
| 1 | $2 \times 10^4$ | $4 \times 10^4$ |
| 2 | $2 \times 10^6$ | $2 \times 10^8$ |
| 3 | $2 \times 10^6$ | $2 \times 10^9$ |
| 4 | $1 \times 10^6$ | $2 \times 10^9$ |
| 5 | $2 \times 10^7$ | $1 \times 10^4$ |
| 6 | $1 \times 10^7$ | $1 \times 10^9$ |
| 7 | $4 \times 10^{10}$ | $2 \times 10^7$ |

---

The endothermic peak temperatures from the first and second runs are, respectively, 130.7°C and 132.7°C (high-density polyethylene [HDPE]). The melting point of HDPE range is 120°C–180°C (248°F to 356°F), while LDPE is 105°C to 115°C (221°F to 239°F) (Table 7.4).

# TABLE 7.4
## Soil Analysis Results

| KM | Quartz- $SiO_2$ | Albite- $NaAlSi_3O_8$ | Microcline- $KAlSi_3O_8$ | Illite- $KAl_2(Si_3Al)O_{10}(OH)_2$ | Chlorite- $(Mg,Fe,Al)_6(Si,Al)_4O_{10}(OH)_8$ | Ca-Montmorillonite- $Ca_{0.2}(Al,Mg)_2Si_4O_{10}(OH)_2 \cdot 4H_2O$ | Palygorskite-$Mg_5$ $(Si,Al)_8O_{20}(OH)_2 \cdot 8H_2O$ | Laumontite-$Ca_4$ $(Al_8Si_{16}O_{48})(H_2O)_{18}$ | Calcite- $CaCO_3$ | Gypsum- $CaSO_4 \cdot 2H_2O$ | Anhydrite- $CaSO_4$ | Halite- $NaCl$ | Hematite- $Fe_2O_3$ |
|------|------|------|------|------|------|------|------|------|------|------|------|------|------|
| 152 | 33 | 21 | 6 | 17 | 8 | | | | 14 | | | | 1 |
| 154 | 22 | 14 | 3 | 40 | 8 | | | | 12 | | | | 1 |
| 172 | 9 | 46 | | 30 | 5 | | | | 9 | | | | 1 |
| 192 | 4 | 1 | | | 4 | | | | 89 | | | | 2 |
| 192 | 17 | 9 | | 33 | 4 | | 10 | | 26 | | | | 1 |
| 212 | 16 | 3 | 2 | 23 | 5 | | 8 | | 43 | | | | |
| 252 | 41 | 15 | 3 | 17 | 5 | | 3 | | 14 | | | | 2 |
| 279 | 32 | 10 | 2 | 36 | 7 | | 3 | | 5 | | | | |
| 297 | 28 | 13 | | 21 | 5 | | 3 | | 14 | 16 | | | |
| 297 | 2 | | | | | | | | 2 | | 1 | 95 | |
| 303 | 47 | 14 | 6 | 3 | 4 | | 2 | | 23 | | | | 1 |
| 303 | 29 | 14 | 2 | 22 | 3 | | | | 16 | 14 | | | |
| 309.6 | 36 | 34 | 3 | 10 | 5 | | | 2 | 8 | | | | 2 |
| 320 | 54 | 14 | 5 | | 6 | | 5 | | 13 | 3 | | | |
| 328 | 26 | 16 | | | 10 | 16 | | | 28 | 3 | | | 1 |

## 7.3  CONCLUSION

1. The polyethylene coating has failed. Mode of failure is cracking, loss of adhesion, and peeling of the PE coating sheets away from the pipe surface. Failure of PE took place in different terrains and different environments and shielded CP current.
2. This pipeline was constructed and coated before the owner took charge of the pipeline. Coating did not follow the owner's standard for PE coating, which calls for a three-coat system; primer, adhesive, and PE.
3. Corrosion was severe at the valleys or low-elevation areas where stormwater gathers, accumulates, and infiltrates to the buried pipe surface, completing the corrosion cell.
4. Liquid-epoxy coating was used to replace polyethylene in the areas where the pipe was rehabilitated.
5. FBE coating on the rest of the pipe does not have major issues compared to the parts that are coated with PE (Figure 7.9).

**FIGURE 7.9**  Refurbished with liquid-epoxy coating.

## 7.4  RECOMMENDATION

1. Conduct pipeline current mapper (PCM) along with DCVG survey along the entire pipeline section (173 km). Based on the findings, generate a plan to recondition the joints having defective coatings, provided that the area is wet and has metal loss history.
2. Rehabilitate as immediate priority the following damaged pipeline sections:
   * from KM 306 to KM 310 (4 KM)
   * from KM 279 to KM 280 (1 KM)

- from KM 296 to KM 298 (2 KM)
- from KM 303 to KM 304 (1 KM)
- from KM 320 to KM 321 (1 KM)

3. Advance the MFL ILI runs for Sections VI and VII from five to two years. Based on the run results reevaluate the other PE-coated pipeline sections.
4. Enhance CP potential levels from KM 164 to KM 184 and from KM 261 to KM 322.
5. Mark each GPS coordinate of the start and end of each reconditioned pipeline section.
6. Update pipeline database to reflect the PE-coated section on safety instruction sheets.
7. Polyethylene coatings should not be used where the pipe service temperature is likely to exceed 65°C.
8. Welded pipe joints can be coated using either heat-shrink polyethylene sleeves or cold-applied, self-adhesive laminate tapes.
9. Field joints should be prepared by wire brushing, although shot blasting is sometimes preferred for sleeves. Heat-shrink sleeves require careful application to ensure a satisfactory bond is obtained.
10. Small coating repairs can be made with either hot-melt polyethylene sticks, polyethylene sheet patches, or tape wrap.
11. Choose the coating material based on the pipeline operating temperature.

# 8 Corrosion Failures
## *Gas Pipeline Explosion*

## 8.1 SITUATION

On Friday, November 22, 2013, the Donghuang II oil pipeline suddenly exploded in Qingdao in eastern China, ripping roads and sidewalks apart, turning cars over, and sending thick black smoke over the city. The blast killed 62 people and injured 136; it was China's deadliest spill since the benzene oil spill in the Songhua River in 2005. The incident led to stoppages in electricity and water in nearby areas. About 18,000 people were evacuated.

The Donghuang II pipeline is owned by China's largest refining company, China Petrochemical Corp., also called Sinopec Group. Qingdao is home to China's fourth-largest port and is popular with travelers seeking seafood and a beach vacation. Sinopec's complex in Qingdao produces liquefied petroleum gas, polypropylene, and styrene, with a total output of more than two million tons a year. Refined products are sold in the north, northeast, and southeastern coastal regions of China, according to Sinopec's website (NACE 2014) (Figure 8.1).

**FIGURE 8.1** Qingdao oil pipeline explosion. (Source: Tong, Shan. 2015. *Cathodic protection.* Training document, Ghana: Sinopec.)

### 8.1.1 EVENTS LEADING TO THE ACCIDENT

The direct cause of the explosion was the ignition of vapors produced from oil leaking from a corroded underground pipeline when workers used a hydraulic hammer that wasn't explosion-proof, resulting in sparks that triggered the blast. In January 2014, Sinopec published a statement on the explosion that blamed worker error and corrosion for the accident; Huang Yi, a spokesman for the State Administration of Work Safety, said that the initial oil leak at the pipeline wasn't properly inspected and that both the pipeline's operator and local government departments bore responsibility for the explosion. Mr. Huang said the initial emergency response was "inadequate" and that workers on the scene failed to detect lingering oil and gas in the eight-hour period between the initial leak and the explosion. Mr. Huang also cited the city's municipal design, saying that the oil pipeline was intertwined with the local sewage system and installed too close to nearby buildings (NACE 2014).

- The pipeline ruptured and leaked for about 15 minutes onto a street and into the sea before it was shut off.
- Hours later, as workers cleaned up the spill, the oil caught fire and exploded in two locations.
- Oil had seeped into underground utility pipes, which could have been a factor in the blasts.
- Oil also caught fire as it spread over the sea.

## 8.2 FINDINGS

- Energy pipeline networks and cities are expanding rapidly, which brings them closer to one another. This provides opportunities for thieves and also leaving lines dangerously close to the general public.
- No nationwide database for pipelines.
- Ineffective investigation procedures. Pipeline investigation was completed in September 2013 but, after the explosion, the effectiveness of the pipeline investigation remains questionable.
- Sinopec completed safety checks across all its facilities and found 8,000 safety problems, ranging from oil and gas well management through standards at rented oil storage facilities.
- Subsequent nationwide pipeline safety check found similar corrosion problems throughout the nation's pipeline network.
- "Safety officials said the blast was caused by sparks from a jackhammer being used to repair a manhole cover following an oil leak. The sparks ignited fumes from the oil that had leaked from a corroded pipe into the city's sewage system."

# References

Ashworth, V., Booker, C. J. L. 1986. *Cathodic Protection: Theory and Practice*. Chichester: Ellis Horwood Ltd.

Baker, M. 2008. *Pipeline corrosion*. Final report, USA: U.S. Department of Transportation.

CalQlata. 2011. Accessed January 17, 2017. http://www.calqlata.com/productpages/00057-he lp.html.

Chalke, P., Hooper, J. 1994. *Addendum to corrosion protection guidelines*. JPK corrosion protection guidelines, United Kingdom: JP Kenny.

Corrosionpedia. 2017. "Cathodic protection and anode backfills." *Corrosionpedia*. September 15. Accessed September 7, 2018. https://www.corrosionpedia.com/cath odic-protection-and-anode-backfills/2/1546.

Corrosionpedia. 2017. *Cathodic protection monitoring*. Accessed January 17, 2017. https://www.corrosionpedia.com/definition/6466/cathodic-protection-monitoring.

CSA. 2015. *Z662-15, Oil and Gas Pipeline Systems*. Canada: Canadian Standard association.

EN12473. 2006. *General Principles of Cathodic Protection*. Switzerland: International Organization for Standardization.

Guyer, J. P. 2014. *An Introduction to Cathodic Protection Inspection and Testing*. Course notes, Stony Point, NY: CED Engineering.

Hale, D. 1984. "Special Report-Slick Way to Increase Capacity." *Pipeline and Gas Journal*, 211 (7) 17–19.

Hilti. 2015. *Corrosion Handbook*. Boston: Addison-Wesley Publishing.

ISO15589-1. 2003. "Inspection and monitoring." In *Petroleum and Natural Gas Industries — Cathodic Protection of Pipeline Transportation Systems - Part 1*, 11–30. Switzerland: IHS.

Javaherdashti, R., Nwaoha, C., Tan, H. 2013. *Corrosion and Materials in the Oil and Gas Industries*. Boca Raton, Florida: CRC Press.

Jorda, R. M. 1966. "Paraffin deposition and prevention in oil wells." *Journal of Petroleum Technology*, 1605–1611.

Klohn, C. H. 1959. *Pipeline Flow Test - Evaluate Cleaning and Internal Coating*. Gas.

Kut, S. 1975. "Internal and external coating of pipelines." *First International Conference on Internal and External Protection of Pipes*. Hertfordshire: University of durham: BHRA fluid engineering. 50–51.

Langill, T. J. 2006. *Corrosion Protection*. Course notes, Iowa: University of Iowa.

Lin, Tian Ran, Guo, Boyun, Song, Shanhong, Ghalambor, Ali , Chacko, Jacob. 2005. *Offshore Pipelines*. USA: Gulf professional publishing.

Lloyd, G. O. 2003. *Atmospheric Corrosion*. Report, UK: National Physical Laboratory.

MESA. 2000. *CP design center*. Accessed March 08, 2017. http://www.cpdesigncenter.com/pages/technical-data/Sacrificial-CP/rev_specs.htm.

MTD. 1990. *Design and Operation Guidance on Cathodic Protection of Offshore Structures, Subsea Installations and Pipeline - Marine Technology Directorate*. London: MTD Limited Publication 90/102.

NACE. 2014. *CORROSION FAILURES: Sinopec Gas Pipeline Explosion*. Accessed September 11, 2018. https://www.nace.org/CORROSION-FAILURES-Sinopec-Ga s-Pipeline-Explosion.aspx.

Natarajan, K. A. 2012. "Advances in Corrosion Engineering." Lecture notes - NPTEL Web Course, Bangalore.

National Research Council (NRC). 1994. "*Improving the Safety of Marine Pipelines*." Washington, DC: National Academies Press. 47. doi:10.17226/2347.

Norsworthy, R. 1996. "High temperature pipeline coatings using Polypropylene over Fusion Bonded Epoxy." *International Pipeline Conference — Volume 1.* Texas- USA: ASME. 253–260.

O'Malley, C. 2018. *High-Temperature Service (Heat-Resistant) Coatings: Industries, Coating Types, Testing Protocols and Consequences of Testing Inconsistencies.* March 16. Accessed September 27, 2018. https://ktauniversity.com/high-temperature-coatings/.

Schmitt, G., Bakalli, M. 2008. *"Advanced Models for Erosion Corrosion and its Mitigation."* John Wiley & Sons, 181–192.

Singh, G., Samdal, O. 1987. "Economics criteria for internal coating of pipelines." *International Conference on the Internal and External Protection of Pipelines.* London, England: University of Durham: BHRA, The Fluid Engineering Centre, Cranfield, Bedford. Paper E1.

Sloan, R. N. 2001. "Pipeline coatings." In *Peabody's Control of Pipeline Corrosion*, by A. W. Peabody, Ed. Ronald L. Bianchetti, 7–21. Houston, Texas: NACE International.

Taiwo, I. A. 2013. *The effect of bitumen coatings on the corrosion of low carbon steels (API 5L X65).* Theisis, Abuja: African University of Science and Technology.

Taylor, D. M. Pipeline Industry, January 1960.

Thomas, L., Burden, T. 2010. *Heavy Oil Drag Reducing Agent (DRA): Increasing Pipeline Deliveries of Heavy Crude Oil.* UK: ConocoPhillips Speciality Products Inc.

Tong, S. 2015. *Cathodic Protection.* Training document, Ghana: Sinopec.

Turkiewicz, A., Brzeszcz, J., Kapusta, P. 2013. "The application of biocides in the oil and gas industry." *Oil & Gas Institute, Krakow,* 103–109.

van der Werff, A. 2006. "The importance of pipeline cleaning: risks, gains, benefits, peace of mind." *Pipeline Technology Conference.* Netherlands: Brenntag Nederland B.V.

von Baeckmann, W., Schwenk, W., Prinz, W. 1997. *Handbook of Cathodic Corrosion Protection.* Houston: Gulf Professional Publishing.

Yong B., Qiang B. 2005. *Subsea pipeline and risers.* London: Elsevier.

Zavenir, D. 2018. *zavenir-blog.* 1 June. Accessed September 22, 2018. http://www.zavenir.com/blog/types-of-corrosion-inhibitors/.

# Key Terms and Definition

**Absorption:** The operation in which one or more constituents in the gas phase are removed to (absorbed into) a liquid solvent.

**Acid Gases:** Impurities in a gas stream usually consisting of $CO_2$, $H_2S$, COS, RSH, and $SO_2$. Most common in natural gas are $CO_2$, $H_2S$, and COS.

**Adsorption:** The process by which gaseous components are adsorbed on solids due to their molecular attraction to the solid surface.

**Anode:** The electrode of an electrochemical cell at which oxidation occurs. Electrons flow away from the anode in the external circuit. Corrosion usually occurs, and metal ions enter solution at the anode.

**Anodic Polarization:** The change of the electrode potential in the noble (positive) direction caused by current across the electrode/electrolyte interface (see Polarization).

**Backfill:** Material placed in a hole to fill the space around the anodes, vent pipe, and buried components of a cathodic protection system.

**Cathode:** The electrode of an electrochemical cell at which reduction is the principal reaction. Electrons flow toward the cathode in the external circuit.

**Cathodic Disbonding:** A process of disbondment of protective coatings from the protected structure (cathode) due to the formation of hydrogen ions over the surface of the protected material (cathode). Cathodic protection systems should be operated so that the structure does not become excessively polarized, since this also promotes disbonding due to excessively negative potentials. Cathodic disbonding occurs rapidly in pipelines that contain hot fluids because the process is accelerated by heat flow.

**Cathodic Polarization:** The change of electrode potential in the active (negative) direction caused by current across the electrode/electrolyte interface. (See Polarization).

**Cathodic Protection:** A technique used to reduce corrosion of a metal surface by making that surface the cathode of an electrochemical cell.

**Coating:** A liquid, liquefiable, or mastic composition that, after application to a surface, is converted into a solid protective, decorative, or functional adherent film.

**Coating Disbondment:** The loss of adhesion between a coating and the pipe surface.

**Corrosion:** The deterioration of a material, usually a metal, that results from a reaction with its environment.

**Condensate:** The liquid formed by the condensation of a vapor or gas; the hydrocarbon liquid separated from natural gas because of changes in temperature and pressure. A stabilized hydrocarbon mixture with a Reid vapor pressure sufficiently low to allow storage at atmospheric pressure (C5+).

**Corrosion Rate:** Because of this process, electric current flows through the interconnection between cathode and anode. The cathodic area is protected from corrosion damage at the expense of the metal, which is consumed at

163

the anode. The amount of metal lost is directly proportional to the current flow. Mild steel is lost at approximately 20 pounds for each ampere flowing for a year.

**Dehydration:**   The process of removing water from gas or liquids.

**Desiccant:**   A substance used in a dehydrator to remove water and moisture. In addition, a material used to remove moisture from the air.

**Dew Point:**   Temperature at any given pressure or the pressure at any given temperature, at which liquid initially condenses from a gas or vapor. It is specifically applied to the temperature at which water vapor starts to condense from a gas mixture (water dew point), or at which hydrocarbons start to condense (hydrocarbon dew point).

**Electrode:**   A conductor used to establish contact with an electrolyte and through which current is transferred to or from an electrolyte.

**Electrolyte:**   A chemical substance containing ions that migrate in an electric field. For the purposes of this book, "electrolyte" refers to the soil or liquid adjacent to and in contact with a buried or submerged metallic piping system, including the moisture and other chemicals contained therein.

**Galvanic Series:**   A list of metals and alloys arranged according to their corrosion potentials in a given environment.

**Gas Processing:**   Separation of constituents from natural gas for the purpose of making salable products and also for treating the residue gas to meet required specifications.

**Ground-Pipe Potential/Pipe-to-Electrolyte Potential:**   Reference potential between pipeline and soil obtained by using a reference electrode in the soil which can be measured with reference to an electrode in contact with the electrolyte.

**Groundbed:**   One or more anodes installed below the earth's surface for supplying cathodic protection.

**Holiday:**   A discontinuity in a protective coating that exposes unprotected surface to the environment.

**IR Drop:**   The voltage across a resistance in accordance with Ohm's Law.

**Instant Switch-Off Potential:**   The ground potential obtained 0.2s ~ 0.5s after a sudden break of the external power or sacrificial anode. As there are no external current flows through the protective structure, the obtained potential is the actual polarization potential without IR drop (voltage drop in the medium).

**Joule-Thomson Effect:**   Change in gas temperature that occurs when the gas is expanded at constant enthalpy from a higher pressure to a lower pressure. The effect for most gas at normal pressure, except hydrogen and helium, is a cooling of the gas.

**Light Hydrocarbons:**   Low molecular weight hydrocarbons such as methane, ethane, propane, and butanes.

**LPG (Liquefied Petroleum Gas):**   Hydrocarbon mixtures in which the main components are propane, isobutane, and normal butane.

**Maximum Protective Potential:**   The maximum allowable negative potential is slightly higher than the potential that can cause cathodic disbonding (usually -1.25V with reference to CSE).

**Minimum Protective Current Density:**   The protective current density required to slow the corrosion to the lowest extent or to stop the corrosion process. The common unit is mA/m$^2$.

**NGL:**   Natural gas liquids are those hydrocarbons liquefied at the surface in field facilities or in gas processing plants. Natural gas liquids include ethane, propane, butanes, and natural gasoline.

**Open-Circuit Potential:**   Potential between the anodes and the reference electrode when disconnecting the anode and the structure. In addition, the difference between the Open/Close Circuit Potential is the drive voltage of the anode.

**Power-Off Potential ($V_{off}$):**   Ground-Pipe Potential obtained at the instant when the cathodic protection current is shut. This value can be regarded as the polarization value since there is no IR drop.

**Power-On Potential ($V_{on}$):**   Ground-Pipe Potential obtained when the cathodic protection current is added, which cannot be used as the protective standards. IR drop must be eliminated when used.

**Polarization:**   The deviation from the corrosion potential of an electrode resulting from the flow of current between the electrode and the electrolyte.

**Polarization Potential:**   The potential across the structure/electrolyte interface that is the sum of the corrosion potential and the cathodic polarization.

**Relative Humidity (RH):**   The amount of moisture in the air compared to what the air can hold at that temperature. When the air cannot hold all the moisture, then it condenses as dew. Relative humidity indicates how moist the air is. A device employed to measure humidity is a hygrometer, while one used to regulate it is called a humidistat or sometimes a hygrostat. Marine environments normally have higher percent relative humidity (%RH) as well as salt-rich aerosols.

**Sour Gas:**   Gas containing undesirable quantities of hydrogen sulfide mercaptans and/or carbon dioxide. It also is used to refer to the feed stream to a sweetening unit.

**Sweet Gas:**   Gas that has no more than the maximum sulfur and/or $CO_2$ content defined by the specifications for the sales gas from a plant; also, the treated gas leaving a sweetening unit.

**Tuberculation:**   The development of small mounds of corrosion products on the inside of iron pipes. These mounds are reddish brown and of various sizes. One solution for tuberculation is biological control. Disinfection strategies may include chlorination, the use of chloramine, or even ozone or ultraviolet light. Disinfectants may kill the bacterial community, but that would not eliminate the deposits. Physical removal of the tuberculation may require a sewer jetter that uses pressurized water to flush out pipes. Another option is to use slip lining to rehabilitate old pipes when other methods prove insufficient.

**Vapor Pressure:**   Vapor pressure specifications may be expressed in terms of a TVP (True Vapor Pressure) or RVP (Reid Vapor Pressure). TVP is a common measure of the volatility of petroleum distillate fuels. RVP is a measure of the volatility of gasoline.

# Index

Milton Keynes UK
Ingram Content Group UK Ltd.
UKHW040054071024
449327UK00019B/559